CAMBRIDGE LIBRARY C

Books of enduring scholarly ꜱ

Botany and Horticulture

Until the nineteenth century, the investigation of natural phenomena, plants and animals was considered either the preserve of elite scholars or a pastime for the leisured upper classes. As increasing academic rigour and systematisation was brought to the study of 'natural history', its subdisciplines were adopted into university curricula, and learned societies (such as the Royal Horticultural Society, founded in 1804) were established to support research in these areas. A related development was strong enthusiasm for exotic garden plants, which resulted in plant collecting expeditions to every corner of the globe, sometimes with tragic consequences. This series includes accounts of some of those expeditions, detailed reference works on the flora of different regions, and practical advice for amateur and professional gardeners.

The Auricula

The plant geneticist Sir Rowland Biffen (1874–1949), who is best remembered for his work on the improvement of English wheat varieties using Mendelian principles, was also a keen botanist and gardener. This short work on the auricula, published posthumously in 1951, contains a full botanical account of the species, but also a social history of this most popular of 'florist's flowers'. Probably introduced to England by refugees from the continent in the late sixteenth century, the auricula, though delicate-looking, is extremely hardy, can be grown in pots, and hybridizes freely, and so it was an ideal plant for competitive growers, especially in the north of England, who in the eighteenth and nineteenth centuries vied with each other to breed ever more spectacular varieties, while adhering to strict guidelines on form and proportion. This work, illustrated with seven black-and-white plates, will be of interest to botanists and garden historians alike.

Cambridge University Press has long been a pioneer in the reissuing of out-of-print titles from its own backlist, producing digital reprints of books that are still sought after by scholars and students but could not be reprinted economically using traditional technology. The Cambridge Library Collection extends this activity to a wider range of books which are still of importance to researchers and professionals, either for the source material they contain, or as landmarks in the history of their academic discipline.

Drawing from the world-renowned collections in the Cambridge University Library and other partner libraries, and guided by the advice of experts in each subject area, Cambridge University Press is using state-of-the-art scanning machines in its own Printing House to capture the content of each book selected for inclusion. The files are processed to give a consistently clear, crisp image, and the books finished to the high quality standard for which the Press is recognised around the world. The latest print-on-demand technology ensures that the books will remain available indefinitely, and that orders for single or multiple copies can quickly be supplied.

The Cambridge Library Collection brings back to life books of enduring scholarly value (including out-of-copyright works originally issued by other publishers) across a wide range of disciplines in the humanities and social sciences and in science and technology.

The Auricula

The Story of a Florist's Flower

Rowland H. Biffen

CAMBRIDGE
UNIVERSITY PRESS

University Printing House, Cambridge, CB2 8BS, United Kingdom

Cambridge University Press is part of the University of Cambridge.

It furthers the University's mission by disseminating knowledge in the pursuit of
education, learning and research at the highest international levels of excellence.

www.cambridge.org
Information on this title: www.cambridge.org/9781108073691

© in this compilation Cambridge University Press 2014

This edition first published 1951
This digitally printed version 2014

ISBN 978-1-108-07369-1 Paperback

THE AURICULA

THE AURICULA

THE STORY OF A
Florist's Flower

BY

ROWLAND H. BIFFEN, F.R.S.

CAMBRIDGE
At the University Press
1951

PUBLISHED BY
THE SYNDICS OF THE CAMBRIDGE UNIVERSITY PRESS
London Office: Bentley House, N.W.1
American Branch: New York

Agents for Canada, India, and Pakistan: Macmillan

Printed in Great Britain at The Carlyle Press, Birmingham

FOREWORD

SIR ROWLAND BIFFEN completed the manuscript
of this book shortly before he died in July 1949
at the age of 75, but he was too ill to give it a
final revision. The Cambridge University Press consequently asked me to do this and to correct the
proofs. Fortunately, Sir Rowland had often discussed
the contents of the book with me, and I had had
frequent opportunities of seeing his fine collection of
Auriculas, including those he had bred himself. On
the other hand, I have no special knowledge of these
plants. The Plates are photographs, taken in 1949
under the author's supervision, of some of the most
interesting Auriculas in his collection.

From boyhood days at Cheltenham Sir Rowland
Biffen was a keen gardener, and he rapidly acquired
an amazing knowledge of garden plants. During his
early career at Cambridge he was one of the pioneers
of Mendelian experimentation, and he subsequently
became famous as a breeder of new wheat varieties.
Little Joss and Yeoman wheats, however, were only
one aspect of a remarkable flair for plant breeding in
general: in his private garden, and constantly assisted
by his wife, new kinds of peas, strawberries, Del-
phiniums, Gladioli, sweet peas and other ornamental
plants were produced, but, above all, Auriculas were
his chief and most constant love. Early in his career
as an Auricula fancier he became intrigued by the

Edged types, and he set himself the task of trying to elucidate their nature and origin. Biffen most happily combined genetics and plant breeding in his techniques, and his perception as an artist of no mean skill also contributed to the creation of flowers of outstanding beauty.

Just before he died Sir Rowland sent some of his choicest Auriculas to the Nurseries, Bartley, Southampton (Mr C. G. Haysom) in the hope that they would be used for further breeding. Others found a home at the Cambridge Botanic Garden, where they will be maintained.

At the end of the book I have included a list of additional references, especially to modern investigations on the Garden Auricula and on the species of Primula from which it has been derived or to which it is closely related.

I am much indebted to the following for assistance in the revision of the manuscript: Mr W. T. Stearn (Lindley Librarian, Royal Horticultural Society), Dr W. B. Turrill (Keeper of the Kew Herbarium), Mr G. Fox Wilson (Royal Horticultural Society's Garden), Dr D. G. Catcheside, Sir William Wright Smith, Mr C. G. Haysom, and Mr R. H. Briggs (Secretary of the Northern Section of the National Auricula and Primula Society).

F. T. BROOKS

Emeritus Professor of Botany
University of Cambridge

BOTANY SCHOOL, CAMBRIDGE
December 1949

CONTENTS

PLATES

PREFACE

SOME THIRTY YEARS AGO I started an investigation
to determine what happened when an Alpine was
crossed with a Green Edged Auricula. It was
carried on to the third hybrid generation, and by
then a large collection of plants had accumulated
which anyone with the instincts of a florist would
unhesitatingly describe as rubbish. This rubbish,
however, was so full of interest and in some cases so
difficult to account for on the generally accepted
principles of heredity that it led me to attempt to
make a more comprehensive study of the Auricula.
In this the somewhat limited point of view of the
florist became a secondary one, and problems involving the origin of the various groups of the Auricula
now in cultivation became the primary one. Con-
currently some problems of cultivation, soon to be
abandoned on account of experimental difficulties
and the impossibility of growing sufficiently large
numbers of plants for the purpose, were enquired
into. By 1939 the investigations were well under
way and several matters on which I could find no
information anywhere were more or less satisfactorily cleared up, whilst others seemed to be on the
way towards a solution. Then the difficulties of
carrying on work of this nature with inadequate and
inefficient garden help necessitated concentration on
the more essential points, with the result that here

and there general impressions rather than conclusive statements can alone be given. Whether they should have been given may be an open question, but the overriding consideration was that, if they were not stated, no opportunity for carrying the investigations further might be available. Furthermore, it may be all to the good to give these views, for some other investigator may be tempted to enquire further into these matters and provide additional information on the parts played by chance and by the efforts of thousands of unknown florists in building up this unique type of plant.

One other consideration has influenced me in making the attempt to piece together the life story of the Auricula. It has been strangely neglected and at only one time in its long history, namely between about 1800 and 1830, were any books published which deal especially with the plant. These were concerned almost entirely with its cultivation, and only fragmentary accounts of other aspects of the subject are to be found in works which are often difficult of access. This field has only partially been explored and there is a possibility that further information is to be found in the horticultural literature of France and Germany. But the story has proved to be more complete than might have been expected and I hope that it may be of interest to those who have fallen under the spell of this curiously fascinating plant.

CAMBRIDGE   R. H. BIFFEN
May 1949

CHAPTER I

THE PLANT AS A WHOLE

W HEN the possibilities of tracing the life story of a plant which has been in cultivation for over three and a half centuries are first considered, it is soon evident that the earliest stages are likely to prove difficult. It is too much to hope that actual specimens will have survived, even in the form of herbarium specimens, for no herbaria were in existence then. Failing these, the literature of the period might be expected to provide some information, though it cannot amount to much, for in those early days books were few and far between. Fortunately, however, the absence of such works is compensated for by the fact that soon after the introduction of the Auricula some of the earliest known botanical books were published. In these great tomes particular attention was given to garden plants, in which respect they differ greatly from modern works on the subject. From them a fairly clear account of the types first brought into cultivation can be obtained, and this moreover can be checked by trying to re-create them from the wild plants subsequently believed to be their parents.

11

These scraps of information, supplemented by pictures and occasional descriptions found here and there in accounts of old-world gardens, diaries, etc., build up, bit by bit, into a consistent whole.

But still more can be learnt from a study of the plant itself. This is of peculiar interest, for the Auricula developed along two distinct courses, one comparable with that of most of our garden plants, the other without any parallel in horticultural history. The result of this second course of development was the production of a flower which, from its earliest days, possessed a beauty and charm entirely its own. Florists took it in hand immediately and from it built up the Show or Stage Auriculas of the present day. They had an unusually fertile field to work upon, for the flower is a complicated one with a number of distinctive features unknown in any other plant in cultivation. At an early stage of their work they defined the 'properties' which they considered the perfect flower should possess, and their ideals have been consistently maintained to the present day. Thus the Auricula grower has to bear in mind a number of conventional 'points', which, owing to the complexity of the flower, is considerably greater than is the case with any of the other florists' flowers. To understand these and also to follow out the course of development of the plant, some knowledge of its general build-up, or Morphology, is essential. This is easily acquired even by those with no interest in botany as a whole.

The plant is a long-lived, somewhat fleshy perennial, which, growing in an untutored state, forms a mat of foliage a foot or so in diameter; it branches repeatedly at about the level of the soil, the branches soon turning upwards and bearing closely-packed foliage, leaves and flowers. The cultivator's ideal, however, is a single-stemmed plant bearing a solitary crown of leaves. The semi-succulent, corrugated stem is about six inches in length with something of the appearance of a diminutive palm tree trunk. But growers, in the interests of tidiness, prefer to bury it at each repotting in order that the crown of foliage may rest neatly on the soil. This underground stem is known colloquially as the 'carrot'. From it new roots and buds, or offsets, are developed.

The inflorescence, or truss as it is universally called, arises from either the apex of the stem, that is, the centre of the rosette of foliage leaves, or from shoots at a lower level. It bears first of all a whorl of leafy bracts varying much in size, some being narrow and pointed and others full-sized foliage leaves, which form a backing to the group of flowers much as a leaf does to the conventional bunch of violets. They are of little significance to the florist but they are often valuable for purposes of identification. Springing from amongst the bracts are the pedicels or footstalks of the individual flowers,

13

commonly known as 'pips'. The number of these, which is almost invariably an odd one, is determined largely by the vigour of the plant, but some sorts of Auriculas normally carry far more than others. Some of the more primitive for instance, may bear fifty or more pips to the truss. As seen at the shows the minimum permissible number is five, whilst trusses carrying up to thirteen are not uncommon. This latter number is often the result of reducing the number of pips, with the object of preventing overlapping and so allowing each pip to display itself properly. Clearly then the length of the pedicel plays a part in determining how many pips shall be allowed to remain, and, other things being equal, a long footstalk has some advantages when a large truss is required,

The flower itself has the characteristic Primula structure with, however, some modifications peculiar to the Auricula., It consists of a tubular portion, the 'tube' (or pipe), within or on which the reproductive portions are situated, and a flattened disc or corolla. The base of the tube surrounds the ovary, which terminates in a slender style bearing at its apex a small knob-like stigma. If the style with its stigma (pointil or pin) reaches the top of the tube the flower is 'pin-eyed'. The stamens (brush, thrum, chives or spices) either form a ring at the top of the tube or about halfway down its interior. In the former case the flower is described as 'thrum-eyed' and its stigma only reaches about halfway up the tube.

The difference in the position of the stamens and stigmas in these two types of flowers facilitates cross-pollination, for if a bee in search of honey visits a short-styled flower it inevitably brushes off pollen from the stamens at the top of the tube, which is then rubbed off on to the stigma of the first long-styled flower it happens to visit. Similarly, the pollen from a pin-eyed flower can be transferred to the stigma, situated half-way down the tube, of a thrum-eyed flower. This, however, does not exhaust the methods of pollination, for pollen from a thrum-eyed flower can easily find its way to the stigma below and bring about self-pollination.

Similar adaptations for cross-pollination are known to be of value to the plants in which they occur, and the cross thrum × pin has been shown in the primrose to produce a better seed crop than crosses between thrums and between pins. But the florists took little notice of this when, about the end of the eighteenth century, they decided that the ideal flower must have a thrum eye, because 'the pin-eyed flower shows a chasm or vacancy very unpleasant to the eye of the curious florist'. So ingrained is this convention that even the hardiest of exhibitors would not dare to include a pin-eyed flower amongst his plants. In fact the conditions for entry of some shows still specifically bar such flowers. In the

15

privacy of the garden however the harshness of the Medes and Persians may be forgotten, for a good flower, even if it is pin-eyed, is beautiful enough to warrant its retention in most collections. Yet in spite of universal and long continued condemnation, pin-eyed flowers seem to be as abundant as ever, and those who believe that by selection they can secure whatever they may wish for must have their faith sorely tried.

The flowering season of the Auricula extends from, roughly, the middle of March until the middle of May with a peak period about the middle of April. In the southern parts of the country it is usually from a week to a fortnight earlier than in the northern. But the actual date at which the plants will be at their best depends largely on the weather conditions earlier in the year, and as show dates have to be fixed months ahead they cannot always coincide with the best period from the exhibitors' point of view. There is often a small-scale, second flowering period in the autumn. Its occurrence is almost certainly to be associated with the climatic conditions of the early summer, for flowers are equally to be found on newly potted plants as well as on those that have been in their pots for two or three years. It is a crop which most growers would willingly dispense with, for even if the flowers expand normally they are often out of character.

Two distinct types of corolla occur in the Auricula.

16

One is that typical of the Primulaceae, the other, whilst showing some general resemblances to it, is unlike anything else to be found in the floral world. The first or normal type has a coloured outer zone (disc, outer rim) surrounding a yellow or white centre or eye (circle, inner rim). The coloured zone is built up of five or, more frequently, six petals, but the number is not necessarily constant, for in a truss in which the first flowers to open are six-petalled the latest are often five-petalled. Each petal is broadly heart-shaped with a notch at its apex, which, however, is not as a rule pronounced enough to spoil the circular outline of the flower. The overlapping of the petals tends also to the building up of a well-rounded pip by doing away with the gaps which produce the starry or windmill-like shape seen so often in primroses and other flowers of the Primulaceae. In addition to this normal flower there is an eight-petalled strain in all of the groups into which florists have divided Auriculas. Little is known about this divergence, but that the extra petals do not mar the outline of the flower is clear from the fact that their presence is neither approved of nor condemned.

The second type of corolla is more complicated. In this the centre is heavily coated with a white layer of meal known as the 'paste'. The outer edge of this is sharply defined by a coloured zone known as the 'body colour'. This does not extend to the margin of the flower but sprays out in small, **irregular streaks** into a green marginal zone which may or may not be powdered with a coating of meal. The diffusion of the body colour into the green margin provides the one touch of informality in the otherwise perfectly symmetrical flower. Flowers built up in this way form the group known as the 'Edged' Auriculas (China-edged, Striped). They lack the central notch seen in the petals of the first type and instead of being heart-shaped they are frequently sharply pointed. The flowers thus have a tendency to be polygonal in outline or, in extreme cases, star-shaped. This is considered to be a serious fault, but if the polygonal shape is not particularly obvious the flower will pass muster with most florists, though it will inevitably be condemned if the corolla is star-shaped. Finally the flower should be as flat as possible, but here again a shallow saucer shape is not altogether unacceptable, although if the depression is marked and the flower consequently cup-shaped, no competent judge will have anything to do with it.

The calyx is the small, green, cup-like structure at the base of the corolla, which, as it is out of sight when the truss is expanded, has received no attention from florists. But there are a number of distinct forms of it which can be made use of for distinguishing between very similar sorts of Auriculas.

There is great diversity in the foliage of the

Auricula, and the shape of the individual leaves cannot be completely described without using the vocabulary botanists have to employ. This, however, is of no great importance to the florist and it will be sufficient to state that the leaf is more or less egg-shaped with the broad end of the egg uppermost. In the broadest forms the shape is rhomboidal and in the narrowest spatula-like. The leaf tapers for about two thirds of its length to its junction with the stem, but has no distinct stalk. The apex, as a rule, is pointed. The blade is thick and fleshy with the midrib and lateral veins so imbedded that the surface is practically smooth. The margin of the leaf may be simple, that is, its outline can be drawn with one sweep of a pencil, or it may be toothed like a saw (snipped, indented). The extent of the serration provides a useful feature for identification purposes and nurserymen often acquire an enviable degree of skill in making use of it.

The most important characteristic of the foliage of the Auricula is provided by the coating of meal found in so many of its sorts. Where this is wanting the leaf has an ordinary, grassy green colouring, and where it is present an exquisite silvery, bluish green. The coating may, however, be so dense that the leaves can be described as white, for the green background is practically obliterated. It is usually distributed uniformly over the surface, but in some sorts it is confined to the margins of the leaf, where

19

it forms a clear-cut edging of silvery white. This is seen at its best on foliage which is otherwise free from meal, but it also occurs in association with a general powdering of meal. Thanks to this meal, which the flattened surfaces of the leaves display to perfection, the Auricula is one of the most beautiful foliage plants in existence, but only those who have access to a large collection can ever know what a surprising range of subtle colour effects are produced by the slight variations in the thickness of the deposit.

When the leaves cease to function, instead of falling they dry up and then generally decay. The position they occupied is indicated by horizontal scars on the corrugated stem. Buds develop on this immediately above the scars, that is, in the leaf axils, and when they have grown to a sufficient size form the offsets by means of which Auriculas are propagated.

There is nothing particularly distinctive about the appearance of the small, rough-coated seeds of the various groups of Auriculas. The most important feature about them is that seed gathered from one particular plant does not necessarily produce similar plants. It is a misfortune which has to be endured and it has made the comparatively slow vegetative reproduction by means of offsets essential when stocks of identical plants are required. This has its repercussion on the problems of seed supply, and

although there are a number of distinct groups of
Auriculas seed is only obtainable under two descriptions, namely Show or Stage and Alpine Auriculas.
In nurseries specialising in their cultivation, the
various groups of Show Auriculas are grown in
houses to which no Alpines are admitted, and vice
versa, with the result that the risks of intercrossing
are reduced to a minimum. Consequently the plants
raised from packets of these two kinds of seed can
reasonably be expected to conform to their description, but the grower requiring, say, Grey-edged
Auriculas, has to be content with picking out a few
from the mixture of Show sorts. A beginner wishing
to acquire a collection speedily may, however, find
this method of distributing seed an advantageous
one.

Only one other feature requires description, and
this, though of no great botanical interest, is of
fundamental importance to the florist. It is concerned with the proportions of several of those parts
of the flower already described above. In no other
plant that I know of has the florist made such an
attempt to solve what is in reality a difficult aesthetic
problem. A century and a half ago he had a clear
idea of what the generality of people considered to
be the most beautiful of the many diverse forms
that the Auricula assumes. Bit by bit he defined the
attributes of a good paste, of a good shape, and so on,
finally crowning his work with a definition of the

proportions which the more obvious features should bear to one another. This is summarized in J. Maddock's *Florist's Directory* (1792) as follows:

The component parts of the pip are the tube (with its stamins and anthers); the eye, and the exterior circle, containing the ground-colour, with its edge or margin: these three should be all well proportioned, which will be the case if the diameter of the tube be one part, the eye three, and the whole pip six or nearly so. . . . The green edge, or margin, is the principal cause of the variegated appearance in this flower; and it should be in proportion to the ground-colour, i.e. about one half of each.

The original standards of perfection are followed nowadays, although perhaps not quite so rigidly as in the past, when the story went round that an essential part of an Auricula judge's equipment was a pair of callipers. Moreover, the more one sees of Show Auriculas the more one is convinced that these standards define the perfect flower with extraordinary accuracy.

Perhaps one of the most surprising features connected with judging is to be found in the readiness with which the uninitiated can distinguish between the good and the bad when confronted with a mixed batch of plants. By way of an example, which may help to encourage any beginner dismayed by the seeming complications described above, the following story may be worth telling. The actors in it were a Welsh miner and myself. He told me that he had worked at the coal face in a mine in the Rhondda

Valley for fifty years, that he had never had the opportunity of growing anything, and that he had never even heard of an Auricula. Standing in front of a group of Alpines he remarked that they were pretty, but it was clear enough that his attention was focused on a batch of Grey-edged plants, good, bad and indifferent, which I had been sorting out and recording. We moved along to them and after gazing at them in silence for several minutes he turned and said to me 'I had no idea that such beautiful flowers could exist'. On the spur of the moment I asked him whether he would pick out the ten plants which pleased him most. He began to do so at once and I can still recall the reverence (that, I think, is the best word to use) with which he handled the plants. The result astonished me, for the selection he made was much the same as I should have made myself. After his departure I worked through it in detail, curious to see what particular features had impressed him most. Of these, proportions came in the forefront, followed closely by outlines; of the remainder I could form no definite opinion. This unplanned experiment, besides confirming the view I had long held that the making of selections was a simple matter, added a fresh interest to Auricula-raising, namely watching the reactions of visitors, and now I can almost welcome even those who lightly pass them over with the remark 'how quaint'.

With a general acquaintance with the build of the

plant, the way is clear for a consideration of the methods of classifying the numerous sorts now in cultivation. The fact that there does not appear to be any reasonably recent account of the subject has made it a puzzling one, especially for beginners. It is easy enough, however, to assign any of the sorts now in cultivation to its proper grouping, but to give it its full name correctly is often a matter which even the best of experts fight shy of. It is the old story of 'too much alike' which gardeners have become far too familiar with nowadays.

The system followed throughout this book can be put shortly in the form of a key:

A. Flowers without a paste	No meal on the foliage or flowers	ALPINE
	Meal present	BORDER
B. Flowers with a paste	Corolla uniformly self-coloured	SELF
	Colour not extending to the edge. Edge green, grey or white	EDGED
	Flowers green on a yellow ground	FANCY

Each of these groups with their subdivisions, where they exist, will be described in chapter IV (p. 62).

CHAPTER II

HISTORY OF THE AURICULA

THERE is an old and probably true tradition that Auriculas were first brought into this country by refugees from the continent during the latter half of the sixteenth century. These settled to a great extent in Lancashire, Yorkshire and the neighbourhood of London—districts which for years to come were the chief centres of Auricula cultivation. By the end of the century the plant was well established in this country, and in its closing years the first English description of it was published in Gerarde's *Herball* (1597). It was then known to botanists as *Auricula ursi*, a translation of the popular name of Beares eares, whilst another name in common use was the Mountain Cowslip. Gerarde described and figured a few plants differing from one another more especially in the colour of their flowers and stated that they grew naturally 'upon the Alpish and Helvetian mountains' and that 'most of them do grow in our London gardens'.

At about the same time Clusius gave a description of the plants in his *Rariorum Plantarum Historia*

(1601).[1] Some years before this he had seen Auriculas growing in the Vienna garden of—to give him Clusius's full description—'C. V. Ioan Aicholztij, Medici und Professoris Viennensis—mei amici veteris', etc. Clusius learnt that the plants grew wild on mountains near Innsbruck. Clusius was then Court Botanist to the Emperor Maximilian II, and one of his special interests was in acclimatizing Alpine plants, especially Primulas, in the great gardens near Vienna. He failed with many of these but grew two *Auricula ursi I* (now *Primula auricula*) and *Auricula ursi II* (probably the hybrid subsequently named *P. pubescens*), successfully enough to be able to distribute them to other gardens on the continent.

Detailed descriptions of the flowers of this period are, naturally enough, scanty. The most comprehensive are those given in John Parkinson's *Paradisi in sole Paradisus terrestris*, published in 1629. In this work the colour, foliage and habit of growth of some twenty sorts are given under the general heading of *Auricula ursi*—'Beares eares'. But the descriptions have to be accepted with caution, for when they were written systematic botany was in an embryonic condition and the lines of demarcation between Auriculas and other species of Primula were not defined. A first reading inevitably leaves the impression that a good many of the sorts described have

---

[1] Before this date Clusius had briefly described six 'species' of *Auricula ursi* in *Rar. aliquoti Stirpium Historia*, 1583.

been lost to cultivation, but this view is discounted to a considerable extent when it is realized that some of the descriptions undoubtedly refer to plants which have no connection with either Auriculas or Primulas. A useful residue remains, however, for Parkinson establishes the fact that the flowers of the early seventeenth century had centres which would now be classified as 'gold' or as 'white', whilst others were 'without any circle at the bottome of the flower'. Thus the two sub-sections into which the Alpine Auriculas are now divided were defined clearly enough at this early date. The third group, the one which lacked the circle, still exists, although it is not often met with.

Parkinson's description of the foliage of his specimen is, for the period, highly detailed, for it takes into account the size and shape of the leaves and the characteristics of their margins. Of more importance, however, for the moment is the fact that he points out that some had mealy leaves and that in others the leaves were 'without any mealiness on them'. Thus, we have one of the earliest indications in print that the Auriculas of the beginning of the seventeenth century belonged to two main groups—one which we now know as 'Alpines' characterized by having no meal, and the other, the 'Borders', in which meal is present in more or less abundance on the foliage.

In a later publication (Parkinson's *Herball* [*Theatrum Botanicum*], 1640), he refers his readers to his earlier account and supplements it with a few more descriptions. One of these, dealing with the Stript Purple Auricula, introduces us to a little-known and long-lost group, namely that of the striped flowers. He describes this as bearing flowers 'variously stript with a kind of whitish-blush colour; some of these will change wholly into one or the other colour, as all or most of the severall sorts of stript flowers, whether *Tulipas*, Gilloflowers [i.e. carnations], etc., are observed often to doe, yet as in them so in these, if they change into the deeper colour, they seldome or never returne to be marked, as they will if they change into the lighter'.

Parkinson's account of the Auriculas of his time is by far the most valuable we have and naturally enough it has been the source from which many writers have drawn their information. In dealing with their compilations it is, however, wise to refer back to the original, for it seems possible for some to read into it meanings which he never intended. Thus a recent writer referring to Parkinson's descriptions states that 'from these it is clear that by this time flowers having a paste centre were in cultivation'.[1] There is no warrant for this. What is clear is that an observer of Parkinson's calibre, who had already

[1] K. C. Cossar, 'Florists' Flowers V. The Auricula,' in *J. Roy. Hort. Soc.*, LXX, 194, 1945.

described the mealy nature of the foliage, would have infallibly called attention to this much more peculiar feature in the flower if it had existed. Logically, too, the compiler, having assumed that the white circle was a paste, should have made the same assumption about the yellow circle and have gone on to point out that yellow-pasted Auriculas were then in cultivation, and so would have given the historians of the future yet another problem to clear up.

Two decades after Parkinson's botanical descriptions we have those of a skilled gardener, Sir Thomas Hanmer. In his *Garden Book* (written in 1659 but not published until 1933), he gives a list of the colours of the Auriculas then in cultivation. He says, 'We have whites, yellows of all sorts, haire colours,[1] orenges, cherry colours, crimson and other reds, violets, purples, murreys,[2] tawneys, olives, cinnamon colours, ash color, dunns and what not?'. He adds, too, a description of two striped sorts, one purple and white, the other purple and yellow. Although the list is admittedly not exhaustive it shows that in the course of a little more than half a century the colour range of the Auricula had become an exceptionally wide one.

Words, however, cannot give a very precise impression of the appearance of the flowers, and for

[1] Haire colour: a contemporary definition of this is light yellow-brown.
[2] Murrey: a dull purple.

this, where possible, one has to fall back upon illustrations. The earliest of these are woodcuts of little value, for the medium is too crude to bring out the subtleties of these flowers, and black and white cannot even hint at their colours. They were soon followed by what are perhaps the finest examples of flower painting which have ever been produced. These are the floral pictures of the Flemish painters of the seventeenth and eighteenth centuries who found the Auricula well worthy of inclusion in the noble masses of flowers they revelled in painting. A good many of their pictures are to be found in the public galleries of this country, and anyone with the time and opportunity at his disposal could probably acquire an almost complete idea of the Auriculas of long ago from a study of continental examples. Mine is based only on the recollections of some forty paintings, for the notes, scribbled for the most part on the backs of envelopes, have disappeared. The outstanding feature of these paintings of Auriculas is that the gold and white-centred flowers were strikingly similar to those to be found in our borders today. The flowers were either flat or goffered like an Elizabethan ruff. This latter feature, which evidently made a great appeal to the gardeners of the Low Countries, is still to be found in plants raised nowadays from commercial seed. The colours, too, were those we are now familiar with, but, owing to the smallness of the sample, the range is not as great as in Hanmer's list. Purple shades predominated and the frequency of quieter shades such as dove-grey, green-grey, café-au-lait brown, etc., indicate that the Border group was extensively cultivated. The flowers were, however, distinctly inferior in the gradation of their colours—a feature known to have shown an appreciable advance in the last half century. Finally the colourings as a whole lacked something of the sumptuousness demanded nowadays. This was almost certainly the case, but it has to be remembered that owing to the shortness of the stems the flowers were generally tucked away towards the base of the bunch, and the painters, concerned as they were with their pictures as a whole, refrained from using their pigments at their greatest brilliancy in the interests of recession. This does not apply, however, to a painting by van Brussel (1754-95) in which two trusses are depicted in full illumination and a third in shadow. They are painted in almost incredible detail, fortunately, for one of them is a red purple striped Auricula. The double Auricula was then in cultivation but no representations of it have yet been found. It may be that this is an indication that they were not easily obtainable, for surely, if they were, no Flemish painter could have resisted the temptation to record their patterning of white or yellow stripes on a coloured ground.

Only a half century or so after its introduction,

31

the Auricula had become well established and it had proved itself to possess most of the characteristics which go to the making of a good border plant. It could be grown apparently on almost any kind of soil, it was indestructibly hardy, not subject to the attacks of those insect pests then worrying the growers of its great rival, the primrose, and with an almost unbelievable range of colours, some of which had a richness and splendour not to be found in any other garden plants. Further, there was to be added to these a feature of the greatest importance. This was that on sowing a pinch of seed there was always the chance of obtaining new sorts, some of which might well prove to be an improvement on any then being grown. This practice was general and plant breeding in its most elementary form was well established. The growers too were happily placed for carrying out these fascinating experiments, for they were men of leisure—the lords of the manor, the squires, the parsons, and so on. The nurserymen of the period had also realized how simple the production of 'novelties' was; from thence onwards the literature abounds with descriptions not only of their wares but also of the plants raised by those amateurs who were willing to dispose of their surpluses. This is of comparatively little interest horticulturally except for the fact that it suggests that attention was focused chiefly on the production of vividly coloured sorts. But it unmistakably establishes the fact that the nurserymen were catering for a wealthy market, for the prices they demanded for the best of their novelties are reminiscent of the days of the Black Tulip boom. One new striped sort, for instance, is said to have been quoted at twenty pounds a plant, and when one considers the value of a pound in those days it seems an extravagance possible only to a very rich enthusiast.

By this time the Auricula had reached a stage in its development at which our fragmentary know-ledge of the development of cultivated plants would suggest that further marked improvements were unlikely to occur. It had run through the not unfamiliar course of producing a large number of distinctively coloured sorts, of giving rise to double and striped sorts, and of increasing in size. But large-scale seedling raising, just as it does today, still offered opportunities for selecting plants showing some improvement on the older types, for instance, in size and colour. It is practically a certainty that this was in progress, but records are scanty in spite of the fact that during the first half of the eighteenth century a number of books on gardening, containing incidental references to Auriculas, were published. The more important of these were R. Bradley's *New Improvements of Planting and Gardening* (1717) and P. Miller's *The Gardener's Dictionary* (1731), which was followed in 1732 by *The Gardener's Kalendar*, the popularity of both of which may be gauged by the fact that they ran through many editions. Their interest is that they give detailed information about the cultivation of Auriculas without, however, providing anything of value about the sorts then being grown. This is unfortunate, for at some time during this period the Auricula started a new and highly distinctive course of development and underwent a series of extraordinary changes which, it is hardly too much to say, converted it into an entirely new plant. These changes are described in detail later and for the time being it is only necessary to say that a new galaxy of characters such as the paste, the edging of green with or without a coating of meal, put in an appearance. The result was a flower having no counterpart in the horticultural world.

Florists seized upon this new type at once and in a short time flowers with green, grey and white edges were being placed on the market. These form the group now appropriately known as Edged Auriculas. As far as known records go, the first of them were the green-edged 'Rule Arbiter' and the white-edged 'Hortaine', both of which were in commerce in 1757. Other sorts followed quickly. The question next arises as to how long it took to work up the stocks of plants then being distributed. Propagation by offsets, the only method of securing a uniform stock, is a notoriously slow operation, so it can safely be assumed that the Auricula started off on its new course of development at least as early as

1750, whilst the number of new sorts in existence soon after this date suggests that nurserymen were busy with this new 'break' at a still earlier date.

As a border plant the Edged Auricula was not a success, for the meal, on which its beauty so much depended, was so fragile that a single raindrop completely ruined it and even a heavy dew was sufficient to convert its almost enamel-like paste into a milky smear. If the flower was to display its newly-found charm properly, protection from the weather was essential. This involved growing the plants in pots under glass and the considerable extra labour such cultivation requires. That the growers thought this worth while was perhaps the greatest tribute to the merits of the earliest Edged flowers they could have paid. It led too, almost inevitably, to the Auricula becoming an exhibitors' plant, and once started on this course it was soon clear that it was in a class of its own with no other plant to rival it.

Descriptions of the Auriculas of the late eighteenth and the early nineteenth centuries are almost entirely confined to the Edged sorts, and from then onwards it is clear that the florists had little interest in the Alpine group though it was still, as now, a popular gardening plant. The number of the Edged varieties also indicate that more and more gardeners were taking up their cultivation. This was indeed the case and the times were ripe for it, for in the early years of the nineteenth century a striking new phase of horticulture was beginning. This was the era of the florist's flower in which men set themselves the task of producing symmetry, an ordered beauty, and refinement in flowers hitherto grown for the sake of their own natural charm. Their efforts were directed not only to the improvement of the Auricula, but to the primrose, polyanthus, carnation, pinks, tulip, ranunculus, etc. The Auricula, already one of the most formal of flowers, had a long start. If it is remembered that the Napoleonic Wars were then in progress and that when peace came in 1815 the country was in a state of intense economic distress, one can only wonder that such a development could have occurred.

Just how far this acted as a stimulus it is impossible to say but certainly the Auricula went on from strength to strength; some measure of its popularity can be found in the fact that during this period three of the best known florists produced books dealing with its cultivation. These were: I. Emmerton, *A Plain and Practical Treatise on the Culture and Management of the Auricula* (1815; second edition, 1819); J. Maddock, *Florist's Directory* (1792); and T. Hogg, *Treatise on the Growth and Culture of the Carnation, Pink, Auricula, etc.* (1820; supplement, 1833).[1] Of these Emmerton's book is the most important, and it has been the quarry from which many subsequent writers have obtained their information.

[1] Hogg's book appeared in several editions.

It is a good source, too, for from his earliest days in his father's nursery and later in his own he had every opportunity of acquiring a sound knowledge of his subject. Moreover he was a research worker, for he kept notes on every experiment and its results. With these three books available for reference one can obtain for the first time a comprehensive view of the position of the Auricula in the horticultural world and also of the hitherto only casually mentioned problems centring on its cultivation. Further, the art of illustration had reached a stage which, at its best, is not equalled today. The Auricula, as was to be expected, received a goodly share of the illustrator's attention and numbers of coloured plates of it were published from time to time. The most useful of these are in *The Florist's Guide* (1827-32), published by R. Sweet. In this forty-two sorts are illustrated and besides descriptions of each there is generally a note on the nurseries from which they were obtainable, together with their prices. Of twenty-six of these no less than twenty-four are Edged flowers and of the remaining two one is undoubtedly, and the other probably, an Alpine. Thus it hardly needs Sweet's comment that the Alpine Auricula was 'not in much esteem by the generality of Florists' to stress the fact that by this time the Edged flowers had swept everything before them. Good as these hand-coloured plates are it has to be recognized that the methods of reproducing the original drawings were not altogether satisfactory, and there are few, if any, illustrations in existence to compare with a single plate depicting two well-known Edged flowers in Grimes's 'Privateer' and Cockup's 'Eclipse', in R. I. Thornton's *Temple of Flora* (1812). Sweet's series is a particularly valuable one, for since nothing like it has appeared in later years it forms the one standard by which comparisons between these old-world and modern sorts can be made. Many contemporary descriptions, often thoroughly critical, are also available to supplement those accompanying the plates. Two opinions exist on this subject. One, frequently held by the older florists of present times, is that our existing flowers are not as good as they were in bygone days. But no man's recollection can go back for over a century, and when pressed they inevitably refer to Sweet's plates as evidence. The second is that the sorts we have nowadays are every bit as good as even the best of the older ones, and those holding this view hint that Sweet's plates represent something of an ideal and are not necessarily an accurate rendering of the plants the artists had before them.

It can almost be assumed that the Auriculas figured were the best available, but something ought to be allowed for the fact that those whose function it is to create beauty are unconsciously biased in this direction and prone to overlook minor blemishes. There is, however, one sound reason for thinking that the plates exaggerate the formality of the flowers: the drawing of their circular or, if seen at an angle, their elliptical outlines is rather mechanical. Be this as it may, it can safely be said that most Auricula raisers would be sorry for themselves if they could not occasionally produce plants good enough to have been included in Sweet's series of illustrations.

The growing popularity of the Auricula resulted in an increasing demand for more and more new sorts. Nurserymen recognised this fact, for Hogg wrote in 1820:

> ... considering the number of years that the Auricula has been cultivated in this country, the varieties are comparatively few: yet from the increasing establishment of Flower Societies, not only in England, but in Scotland and Ireland also, in which Societies silver cups, and other prizes are yearly awarded to those members who exhibit the finest and most perfect flowers, and from the great pains and attention now paid to the raising of seedlings, we may fairly expect, in the course of a few years more, a very considerable accession of new flowers.

Yet 'comparatively few' is a modest statement, for in the second edition of Emmerton's book a list of ninety-four sorts is given, and in the copy from which this number is taken the names of another nine have been added, apparently by a contemporary nurseryman, since alongside each are numbers showing how many he had in stock. Emmerton also recognized that his list was by no means complete and drew attention to the fact that many other sorts were in the hands of the north-country nurserymen. In all probability an estimate made about this time that about two hundred sorts were then being grown was not very wide of the mark.

Prices ranged from 7s. 6d. per plant for the older sorts up to two or two and a half guineas for the newer ones, but their raisers, often amateurs, frequently received only from five to ten guineas for the limited stocks they had for disposal. On the continent, too, the position was much the same as in this country. Contemporary French accounts point to the widespread cultivation of the plant often on a very considerable scale, as is indicated by the fact that, for the proper display of their treasures, the monks of the Abbey of Tournai required no fewer than fifteen Auricula theatres or stages. In Holland also many new sorts were being raised, some of which found a ready market here, whilst in Germany there were so many that F. A. Kannegiesser (*Aurikel Flora nach der Natur gemahlt*, 1801) was able to figure one hundred and forty-four of them.

By 1820 Auricula growing had become a fashion, and in a curious if rather indefinable way it fitted well into the decorative schemes so characteristic of the Regency period. As a fashionable flower too it had to be displayed to neighbours and friends, and the custom of building special theatres or stages for the purpose grew up. Thus the still persisting name of 'stage' auriculas came into existence. Growing in frames, the general practice—and an excellent one from the cultural point of view—had its drawbacks, for the plants towards the centre were rather inaccessible and inspection at eye-level was essential for the full appreciation of the individual truss. The stages to which the plants were transferred from the frames as soon as they came into flower were permanent wooden structures open to the north, with a water-tight roof and often with ventilating shutters at the sides. The back was commonly painted with some dark colour or, if the owner's decorative sense matched the times, with a landscape view, or again temporary curtains of richly coloured velvet formed the background. As a final refinement, mirrors were attached to the two sides to give the impression of greater size to the exhibit. Within this structure two or three shelves arranged staircase-fashion formed the support for the plants. Then, when the flowerpots had been painted a quiet shade of green to get rid of their somewhat discordant red colouring, came the difficult task of arranging the plants to the best advantage.

The stage was something more than a quaint conceit; it was an admirable device not only for giving friends an opportunity to see exquisite flowers under good conditions but also to afford the more knowledgeable ones a chance of going over them point by point.

41

About 1820 competition between growers became general and flower-shows came into existence on a scale which nowadays seems to be remarkable. They continued through all the changes of the times until the middle of the nineteenth century, so long, in fact, that some of the older growers of today can recall being told about them in their younger days by men who had taken part in them. These shows were particularly widespread in the neighbourhood of London and in Lancashire, Cheshire and the southern parts of Yorkshire, and it is hardly an exaggeration to say that every village had its own. They were very different from the village shows of today, for they were held in nurseries, private houses, and especially in the North in the local public-houses. In the wealthier neighbourhood of London, local patrons provided prizes of a piece of plate or of sums of money varying from two to ten guineas. In the North the prizes were more modest and distinctly utilitarian, a copper kettle being apparently one of the most appreciated. Difficulties of transport were great and this more than anything else limited the size of each man's exhibit to a pair of plants. Small though they were, the quality of these shows must have been wonderful, and judges must have found it hard to discriminate between the exhibits of growers who brought in the very pick of a collection of 200 or 300 plants. These larger-scale growers tended to become pot hunters, for the possession of an unbeatable pair of plants made it easy for them to tour the district and collect all the prizes on offer. In the village shows of the North, however, exhibiting was a more homely procedure and those taking part in it could sit round the table and, with their plants and a mug of beer before them, discuss, as Auricula growers have always done, the good and bad features of all that were on show.

Unfortunately, the day of these shows has gone and the exhibitors' needs have to be met by the shows at Manchester and London respectively of the northern and southern sections of the National Auricula and Primula Society. The limitation of these shows to two places has brought in a difficulty, again of transport, for only the most enthusiastic of growers at any distance from either of these centres can be expected to make an effort to exhibit. Thus the social value of the shows of the past has greatly diminished. This is offset, however, by the fact that the larger commercial growers are given an opportunity of bringing their wares to the notice of a wider public. They avail themselves of it effectively and the large number of plants they display provides everyone present at the shows with an unrivalled chance of seeing well-grown examples of plants of all of the various sections, and of realizing something of the great diversity of sorts still in cultivation.

This paragraph, however, anticipates events and a return has to be made to the early years of the nineteenth century. Then the growing of Auriculas was mainly the hobby of the well-to-do, but their cultivation was steadily increasing amongst those less well endowed and it was clear that the Auricula was on the way to becoming everyone's flower. No class of workers took up the task of growing the many exquisite sorts then being introduced with more enthusiasm than the silk weavers of Lancashire and the neighbouring counties, and in the course of a generation they brought Auricula cultivation to a pitch which has never been equalled. This was an interesting development and one difficult to account for completely. An unusual appreciation of the subtle charm of the flower could, perhaps, be expected from those continually handling a material as beautiful as silk, and further, the nature of their occupation gave them better opportunities for attending to their plants than fell to the lot of most manual workers. Theirs was essentially a home industry and the days of factory production, though looming ahead, had not yet arrived. As a consequence they could, whenever they wished, still the click of their looms and assume the more congenial occupation of inspecting the contents of their makeshift frames. This seems a plausible explanation, but contemporary accounts include miners among the enthusiastic Auricula cultivators, and it is a far cry from the bottom of a mine-shaft to the loom at home and a nearby garden. But it is not a problem to worry over and one can be content with the pleasing mental picture of the monotonies of a homely industry being broken by efforts to raise flowers of surpassing beauty.

The changing economic conditions and the changing standards of taste of early Victorian days may have had an effect on the cultivation of Auriculas. Whether this was so or not, it is certa'n that the wave of enthusiasm which had carried the north-country growers so far was noticeably subsiding. The crest had been reached, as far as such occurrences can be dated, about 1850, and by 1870 there was every sign that it had passed, and that the Auricula no longer retained its almost amazing popularity. The gap between 1870 and the present might conceivably be bridged by some centenarian but there is little likelihood of our ever hearing a first-hand account of the position in his early days. Many present-day growers, however, can recall the stories told them by their fathers of the splendour of those days. These are frequently the men mentioned earlier who insist, in spite of any real evidence, that the Auricula has deteriorated since then.

Waning popularity, however, had no effect upon the output of new sorts, and the names of the most prominent raisers of the later years of the nineteenth century—Lightbody, Barlow, Horner and Ben Simonite—are still household words to modern growers. Ben Simonite, the best known of them, was responsible for the introduction of no fewer than

45

seven sorts in the year 1904. Some of the introductions made soon after 1870 still survive. None of them appear to be obtainable in the open market, but they are lovingly cared for by the few amateurs who possess them. These sometimes share with the public their own pleasures by providing photographs for reproduction, for the most part, in the weekly gardening papers. But to judge of their merits is almost impossible, for the Auricula is difficult to photograph satisfactorily, and even if its fine details are secured they are lost during the processes of reproduction.

The raising of new sorts was not confined to the Edged group, and for the first time for many years a sustained attempt was made by Turner of Slough to improve the Alpine group. He had already raised and distributed several outstanding Edged sorts, one or more of which are still being grown, when he set out to make the Alpine group a florists' flower. His aim was to obtain richer colour effects in both the uniformly coloured and the gradated flowers as well as more pleasing proportions between the coloured zone and the centre. In this he met with a great measure of success, and gradually the older view that the Alpines were unworthy of exhibition became obsolete. A few special classes for the group found their way into the schedules and of late their number has increased to meet an unquestionable demand. Nowadays these classes have become popular and they attract the attention of many visitors who disclaim any interest in the more formal Show flowers.

A number of coloured plates of Turner's new sorts was published in *The Floral Magazine* (1861-71), and though their merit is not great, they are sufficiently good for one to be reasonably certain that some of them are still surviving in our borders.

At about the same time James Douglas of Great Bookham was devoting the efforts of a lifetime to further popularize the Auricula. Not content with breeding plants of both the Edged and Alpine groups on an unprecedented scale, he did all he could to provide the younger generation of horticulturists with authoritative information on every theme connected with the Auricula. He helped too in the organization of shows at which he put up exhibits such as have never been seen since. But his most distinctive contribution was to place at the disposal of members of the National Auricula and Primula Society, free of cost, seed from his incomparable collection.

More recently, the Bartley nurseries, near Southampton, began to specialize in raising and distributing new Auriculas. After the death of their original owner, G. H. Dalrymple, whose knowledge of both florists' and garden flowers was unusually extensive, the work was taken over by C. G. Haysom. He has already produced a considerable number of first-class sorts, especially in the Edged group, and as his declared object is to improve the constitution of the plant, one may hope that still better varieties are in the making.

That, then, is the position today. What the future may be is anyone's guess. The optimistic may point out that during and after the Napoleonic Wars the Auricula achieved a surprising popularity, and go on to say that history repeats itself. The pessimistic may insist on the troubled conditions of the day and the difficulty of finding the means and the leisure for cultivating the plant. For him there is no hope that it will recover its former greatness. But surely, however badly things may go, there will always be many who can find their happiness in contemplating the mysterious haunting beauty of this, the most perfect of all the flowers which human efforts have brought into existence.

CHAPTER III

MEAL AND COLOUR

THE outstanding feature which differentiates the Auricula from all other flowering plants, and at the same time plays an overwhelming part in building up the curious and mysterious beauty of the edged flowers, is to be found in the presence of meal in the centres and often on the marginal portions of the corolla. Thanks, too, to its prevalence on the leaves the Auricula is one of the most beautiful foliage plants in existence. Far too little attention has been given to the meal in the past, and both the old and new descriptions of plants, all too frequently, only refer to them as being either mealy or mealless.

Before one can make a detailed study of the structure and distribution of the meal it is necessary to know something about its occurrence in the wild-growing ancestors of the Auricula. A fuller account of these is given in a subsequent chapter and for the moment it need only be stated that there are two Alpine species, one, *Primula auricula*, which has mealy foliage, and the other *P. hirsuta* with grassy green leaves on which no trace of meal can be detected.

The meal consists of an unusual type of glandular hair which occurs in many species of Primula. Each hair, when examined under a microscope, is seen to be a transparent globe carried on a short stalk composed of smaller cells. Protruding through the thin walls of the globe are large numbers of almost incredibly fine filaments of a waxy or resinous nature. The whole structure thus has much the appearance of a short-handled household mop. The globular portion is too small to be visible to the unaided eye of most persons, but the filaments make the whole hair just visible to the naked eye as a minute, silvery spangle, provided it is well isolated from its neighbours. More commonly, however, the hairs are massed together into a miniature forest in which the branches are represented by the filaments. Then the meal forms a more or less uniform white layer, generally smooth enough in appearance to the naked eye, but to an insect walking over it, it has a soft, rough surface on which foot-prints show up like those of a rabbit in the snow.

There are two kinds of these glandular hairs differing only, as far as can be determined, in their size. The larger give the surface on which they are borne a frosted appearance, but the close matting together of the smaller hairs results in the formation of a smooth, non-sparkling layer with a matt or, in some lights, an enamel-like appearance. This is seen at its best in the paste in which every cell of the epidermis seems to give rise to one of the smaller hairs. It is found also on the leaves of many sorts of Auriculas, the rosettes then having an artificial appearance, suggesting that they have been carved out of a block of milk-white jade by a master craftsman.

The amount of meal developed on the leaves or on parts of the flowers does much to determine their general appearance. Where there is no meal on the foliage, as for instance in the Alpine group, the plant has a somewhat commonplace look and almost completely lacks the attractiveness of the mealy sorts. Where meal is present the plant becomes a thing of beauty on account of the subtlety of its colouring. A thin deposit of meal gives the foliage a soft silvery green tint, a slightly heavier one makes it a blue-green, and, as it increases, the colouring becomes more and more grey-green until, when the maximum density is reached and the green of the background is entirely masked, it is an almost pure white. When seen at its best, as the leaves begin to fold back in the early spring and the sunshine brings out the sparkle of the meal, one may wonder whether the Auricula would not have found a place in our gardens even had it been a flowerless plant.

The meal is not necessarily always white, and a greyish tint, condemned in no uncertain terms by all florists, is not uncommon in the paste of some flowers. No excellence in other respects is sufficient to prevent these from being consigned to the waste heap. There is also a pale sulphur yellow colour, so little known that it does not seem to have been recorded, which cannot be condemned out of hand, even if the fiat has gone out that the paste must be white. This soft yellow colouring, when associated with the dark body colour and the bright green of an edged flower, gives a colour scheme of considerable beauty. Were it substituted for the white of a white-edged flower the result would be an Auricula flower of, presumably, unusual attractiveness. The operation of creating such a flower should not be a difficult one, and anyone interested in it will find a useful source of this colour in *Primula auricula Bauhini*. The second generation of such crosses would further provide a number of plants with leaves powdered with this yellow meal, so producing a still wider range of foliage colours.

The distribution of the meal in Auriculas presents some interesting features. In the mealy parent (*P. auricula*) it often covers the whole plant except for the coloured portion of the flowers, and the coating is uniform. This is true of many sorts in the Border and Self groups, but in others the distribution is by no means general. In the Alpine group, almost always described as meal-less, the foliage is definitely so, but the main flowering stem, the pedicels of the flower, and the calyx are, perhaps more often than not, coated with a more or less dense layer of meal. The same state of affairs is found in the Green-edged group. Further, in the Grey- and White-edged, the Fancy and Self groups, all potential meal producers, since their flowers carry a paste, the foliage may or may not be mealy. At present no more than a recognition of the facts is possible, and an explanation from some future geneticist must be awaited.

The fact that the seedlings of all Auriculas have a grassy green colour is well known to all who have raised them, and it is generally recognised that it is only comparatively late in life that those which will ultimately develop into mealy sorts acquire this characteristic. This generally happens after about nine months growth but it may take place as much as a year after the seedlings have appeared above ground. Then a light coating of meal is to be found towards the base of one or more of the leaves, and after that every leaf the rosette produces carries its predestined crop of meal. But when buds are formed below the point on the stem at which meal formation started, the leaves they give rise to are as free from meal as the original seedling. In due course the leaves develop meal but for some months the plants produce two distinct types of foliage, and as it is these buds which grow into off-sets there arises the curious state of affairs that meal-bearing plants are propagated from meal-free tissues. This continues throughout the life of the plant, the growth from the apex being always mealy and that from the base meal-less. One consequence of this has been a puzzle to generations of Auricula growers. Hardly a season goes by without one of them sending to one of the weekly gardening papers a photograph of a plant bearing a truss of green-edged and a truss of grey-edged flowers, usually with the question 'which group does this plant belong to ?'. In the dozen or so of plants which have come under my observation the green-edged has been a basal, and the grey-edged an apical truss; there are no reasons for assuming that this is not always the case. If this is granted the explanation is evident. The formation of meal has again been delayed, just as it is in the offsets, and the plant is really a member of the Grey-edged group. A direct proof of this is possible by propagating the 'freak'. This is not quite as easy as rooting down normal offsets, for the base of the truss springs straight from the stem, and the foliage a bud should have is represented by small almost bractlike leaves. But by slicing off the portion of the stem carrying these, one was rooted and nursed on into a plant which bore, not green, but grey-edged flowers.

The occurrence of these two kinds of flowers on the same plant provides an opportunity for making a precise comparison between them, for, as so rarely happens, it is certain that they have grown under identical environmental conditions. When this is done, the one difference found between them, apart from that of mealiness, is that the green flower is distinctly larger than the grey. The number of plants is too small to base a sweeping generalization on, but it tends to confirm the impression that the flowers of the green group are generally larger than those of the grey and that these in turn are larger than those of the white group. In order to prove, or disprove this, a lengthy series of measurements and a statistical examination of the resulting figures would be necessary. Assuming, however, that this view is correct it would appear that the production of meal inhibits to some extent the growth of the tissues bearing it. The occurrence of 'mousey' flowers provides some evidence that this hypothesis is a workable one. The description 'mousey' is given to flowers of the Edged groups which have an abnormal, ragged outline, suggesting that at some stage of their growth they had been nibbled by a mouse. At the bottom of each nibble an edging of meal is to be found whilst the margins of the out-growing portions, which make the flower ragged, are meal-free. Such a result would be the outcome of a broken mealy margin, for growth would have been checked at the points where meal was being developed and unchecked in others.

The colour range of the Auricula, which baffles the descriptive powers even of the writers of cata-logues, is as wide as, if not wider than, that of any plant in cultivation, and thus it becomes a matter of some interest to those who grow it to discover how this great diversity came into existence. On the face of it the problem seems as if it must be one of more than ordinary complexity but it has turned out to be a comparatively simple one, and the broad outlines, if not all of the minor details, are now reasonably clear.

As a basis for this examination it is obviously necessary to know what colour contributions were made by the two parent species, *Primula auricula* and *P. hirsuta*. In the first of these a clear imperial yellow is characteristic of the type and of most of its recognized variants, in the second the colour is usually described as rose or rosy carmine. But though rose is a fair description for the majority of the flowers of *P. hirsuta*, it is not universally applicable, for the colour is very variable, and it is easy to pick out a series in which at one extreme a deep carmine or beetroot colour and at the other almost pure white find a place.

The starting-point then is two colours, yellow and rose, plus a capacity to give rise to an indefinite series of tones of the latter. It seems an inadequate one, for no mixture of these two pigments by the hands of the most skilful of painters could produce more than a fraction of the colours we are familiar with. Thus it has come about that many consider that the parentage of the Auricula as given above is incorrect.

The colouring matter of flowers has not, however, so simple a composition as that of the pigments the painter uses. That of *Primula hirsuta* has been isolated and its properties extensively investigated. To the chemist it is known as 'hirsutin', and he considers it to be very similar in constitution to anthocyanin, that is, the widely distributed purple colour seen at its best in the pickling cabbage. The property which concerns us most is that its colour is dependent on the medium in which it is dissolved. If the medium is neutral the colour is purple, if it is acid crimson, and if alkaline blue. In this respect then it resembles litmus, the indicator in general use to determine whether a solution is acid or alkaline.

Thus the rosy colour of *P. hirsuta* is capable of transformation into two other colours by slight changes in the acidity or alkalinity of the cell sap, and what has been considered as a single colour is, depending on conditions, a crimson, a purple (rosy) or a blue. (The names here given to the colours are but a rough approximation. They are sufficient for the purpose in view and have the advantage of obviating the attempt to distinguish between crimson and carmine, mauve and lavender, etc.) Further, as the wild plant indicates, there may be an almost infinite series of tones of each of the colours. Crimson may be anything from almost black to pale pink, purple may intensify to black or dilute down to mauve, and blue may become a pale lavender. The palette then is not nearly as limited as it appeared to be at first sight.

The second parental colour, yellow, shows none of these complications although there are grounds for thinking that there is more than one tone of it. This view is strengthened by the existence of a beautiful white *Primula auricula* in cultivation which may have played some part in the production of buff colours.

A study of the distribution of these pigments in the flower carries the story much further. In *Primula auricula* the colour is in the form of glistening yellow particles uniformly distributed throughout the tissues of the corolla, but in *P. hirsuta* it is in solution in the cell sap and confined to a specialized layer of cells. These take the place of the usual epidermal cells. They are easy to examine, for with a little care small sheets of them can be stripped off and mounted outside uppermost, in a drop of water for inspection under a microscope. Looking down upon the preparation one is reminded of a layer of pears in a fruit tray, each pear resting on its broad eye end with the narrow stalk end standing up in the air. These pear-shaped cells have a transparent outer wall thickened at the stalk-end to form a miniature lens. Each one is filled with cell sap in which is dissolved the pigment hirsutin. The concentration of the pigment is far from constant, for if the layer has been taken from a gradated flower,

58

then at one end of it the cells appear almost black and as the slide is moved towards the other end the colour becomes paler and paler. In a uniformly coloured flower, however, the colour of the cell sap solution is practically constant. The tissue underlying the pigment layer in *Primula hirsuta* is colourless or white.

The immediate result of crossing the two wild-growing species was the production of a hybrid in which the colour layer was developed on the yellow ground of *Primula auricula*, with the result that its flower had a colour resembling that of burnt sienna. A new factor thus comes into consideration—the dual nature of the colouring of the Auricula.

It is clear that this has at once doubled the colour range, for a transparent wash of hirsutin super-imposed on a yellow background produces a colour very different from what it would be on a white ground. Anyone used to handling water colours will realize at once what the result would be, and anyone attempting to describe it in detail would quickly give up the attempt. All that can usefully be said is that crimson overlying yellow gives scarlet, purple on yellow gives brown, and blue on yellow gives maroon. If, however, a clearer understanding is required it can easily be obtained by constructing a simple colour chart by laying washes of the appropriate shades of crimson, purple and blue, gradated from the deepest to the palest tones, on strips of yellow and of white paper. This also provides a useful method for recording purposes if the strips are marked off into numbered areas. The colour can then be matched against the right scale and recorded as a number instead of attempting to set it down in words.

This laying of one colour over another produces its most interesting effects in the gradated flowers which have a yellow background, for then the superimposed colour alone becomes lighter from the centre outwards whilst the yellow ground remains constant. Thus an intense crimson, deep enough to obscure the yellow near the centre of the flower, becomes, as it tones down, a coppery scarlet, an orange, and at the margin of the flower an almost pure yellow colour. It is to this especially that Alpine Auriculas owe their extreme brilliancy of colouring. Its velvety richness too may be due to an unexpected phenomenon associated with the pear-shaped colour cells. If a rose-coloured flower is held at an angle to the light and examined under a hand lens with a magnification of 16 or so, regularly shaped points of a brilliant crimson colour are seen against a uniformly dark background. The appearance is, on a small scale, identical with that seen when a car's head-lamps light up the warning signs along the roadside. It is due to the same cause. The thickened end of the cells, already described as lens-shaped, act as lenses and bring whatever rays of light reach them to a sharp focus in the same manner as the faceted lenses of the road signs. The colour of the flower, then, is not due to a flat and uniform colour-wash but to myriads of minute, intensely illuminated points.

Both colour cells and the glandular hairs of the meal are developed from epidermal cells, and, as a rule, a patch of the epidermis gives rise to either one or the other if and when it departs from its normal structure. Very occasionally, however, a wayward cell goes its own way and produces a hair which then appears as a minute white dot in the coloured area. This fault cannot occur in the meal-free flowers of the Alpines, but it is common in the body colour of the Edged group, particularly in the neighbourhood of the paste. It may be corrected easily in flowers wanted for exhibition by lightly touching the hair with a small brush damped with alcohol. The filaments then dissolve immediately and, as the naked eye cannot detect the globular portion, the fault disappears.

CHAPTER IV

THE GROUPS OF AURICULAS

THE large number of sorts of Auriculas in cultivation, even in the earliest days of the plant's development, made some system of classification essential, especially for those who were concerned with its sale and distribution. Such a system came into existence slowly and even as late as the beginning of the nineteenth century one of the best known florists could write 'there are two classes of Auriculas, the plain and the striped'. The lack of a system of classification was unfortunate, particularly for those interested in the plant's history, for when reading descriptions of the old-world sorts, it is often impossible to decide which group a plant belongs to. Thus endless confusion arises between those now distinguished as Alpines and Selfs, and matters are made worse by the use of the term 'striped', quoted above, which was commonly used for both typically striped flowers and for the highly distinctive Edged group.

The system of classification in general use now was devised, primarily, in the interests of exhibitors and judges: broadly speaking, it is a satisfactory one.

Its simplicity is a great merit and the reasonably sharp definition of the various groups makes its application easy. But, as with all plants subjected to long periods of hybridization, absolute precision cannot be claimed for it.

The simplest group is that known as the Alpine. Its distinguishing feature is the complete lack of meal on the foliage and corolla. The flowers throughout the group are large, larger in fact than in any other group. The corolla is generally six-petalled but the more primitive five-petalled sorts are still abundant. A distinctive eight-petalled flower is also commonly met with. The flower is either flat or fluted. The centre of the corolla, the eye, is either white or yellow in colour owing to the absence of any other pigment layer in this area. This feature is made use of to divide the group into two sub-groups —the gold-centred and the white-centred—the term 'gold' being used rather widely to cover a range of tints varying from buttercup to sulphur yellow. The petals are heart-shaped with a single notch at the apex and fused together at the edge of the centre. In the older Auriculas there was a gap between each of the petals, and the flower was not inaptly described as windmill-shaped. This feature has long since disappeared. The petals now overlap and so give the flower a clean circular outline. The overlapping is not haphazard, for the left portion of the petal lies over the right portion of its neighbour except in one petal in which both margins are exposed. This break in the symmetry of the flower is occasionally overlooked by draughtsmen, and floral decorations of china and porcelain often suggest that either they had no specimens before them or that this detail was not worth worrying about. The many colours occurring in the petals of Alpine Auriculas are as rich and sumptuous as colour can be, and the flower has a velvety lustre which adds greatly to the splendour of the general effect. The colour may be either uniform from the centre of the flower to its margin or it may be gradated, the deepest tones then being at the centre and the palest towards the edge. This gradation, when associated with a pronounced overlapping, gives the flower the appearance of a Catherine wheel, for the light-coloured margins of the petals stand out in sharp relief from the deeper tones of the inner portions of the colour zone like curved spokes.

The formation of a good collection of Alpines is an easy matter. As a nucleus, plants of the finest sorts can be bought at reasonable prices, and by using seed saved from these the collection can be rapidly increased. Failing this, seed of excellent quality can be obtained from any firm which has specialized in Auricula cultivation. From a packet or two of seed from a reliable source it is not unreasonable to expect a crop of plants, many, if not most, of which will be of first class quality. In selecting these, any plants showing defects should be discarded and every effort directed towards keeping up the rare standard of perfection which has now been attained. The commonest defect, often particularly obvious in cottage gardens in which the plants may have been grown for generations, is an excess of coarse, ragged foliage. A neat rosette is much to be preferred. A weakly stem, incapable of supporting the head of flowers under bad weather conditions, is another feature which need not be neglected even if, for show purposes, tying the truss to a light support is the prevalent custom. Too many flowers to the truss, a feature which cannot always be gauged in a plant flowering for the first time, is undesirable. Huddled together in a solid, rumpled mass the flowers have an unpleasing appearance even in the open border. The ideal is that each pip should just touch its neighbour, for only then can its shape and colouring be appreciated.

Once the undesirable plants have been rejected the raiser is at liberty to make a purely personal choice as to what should be saved; for as regards the Alpine section, florists have not, so far, provided any clear-cut specification of what, in their opinion, the ideal flower should be. The features he will almost certainly keep in mind are brilliancy of colouring and size, but he may be doubtful about the respective merits of the flat and the fluted flowers. If he has any of the instincts of the florist he will discard every plant bearing flowers which are not completely flat, but if the ladies of the household are assisting in the work of selection a plea may be put in for their inclusion. The one serious difficulty which will arise is in connection with the size of the eye. Many would agree that in some of the sorts nowadays, especially those with a coppery orange colouring, the eye is far too prominent. The result of this is that the plants have a spotted appearance and the true colour is not as obvious as it should be. Without attempting to define the most desirable proportions it may be said that a neat, well defined and non-aggressive eye is desirable. Somewhat rarely plants may be found in which the eye is almost non-existent but which nevertheless have quite attractive flowers.

The Alpine Auriculas, although generally grown as border plants, can be made use of in other directions. Many find them valuable in the rock garden, whatever purists may say about their inclusion in it. The site chosen should not get the full glare of the midday sun, and if it is provided with underground irrigation the plants may be relied on to be flourishing when many of their real Alpine neighbours have disappeared. Failing this, the routine watering of the rock garden is all that is required. Another use for the Alpines, and one that will probably increase, is as decorative plants for conservatories and cool green-houses. Nowadays vanishing incomes and

66

labour and fuel difficulties make gardening problems a nightmare to many. If a green-house, stripped of its heating apparatus, can be kept in commission Alpine Auriculas can be grown in it to perfection with very little trouble. When the plants have finished flowering, they can be put out in the open, where almost all the attention they require is an occasional watering, and then returned to their flowering quarters when the winter comes on.

The Border Auriculas are contemporary with the Alpines but the older writers did not distinguish between the two groups, with the consequence that the story of their development cannot be traced in any detail. Even today the differences are often overlooked and a great deal of unnecessary confusion exists.

A sharp line of demarcation is provided by the fact that the Borders have a coating of meal on their foliage and flowers. But setting that on one side they are still not precise counterparts of the Alpines; for they show in addition a number of minor though distinctive differences. The degree of mealiness on the foliage varies considerably and even when rain has washed much of it away there is generally sufficient remaining to give it a silvery grey lustre. On the flowers the coating is confined to the eye. It is generally the merest of light powderings except in the immediate neighbourhood of the throat, where it may be dense enough to form a distinct ring. Even under bad weather conditions its disintegration leads to no disfigurement of the flowers.

Again, the colouring of the Borders is almost distinctive enough to distinguish them from Alpines at a glance. The ground colour of the flowers may be either yellow, white, or buff whilst the superimposed colours are those of the Alpine group but markedly lighter in tone. Thus, taken as a whole, the Borders have a subdued colour effect. This is especially so where the very pale tones of the superficial pigment cover the buff ground. A pale purple then gives rise to an unpleasant muddy grey, and if still paler, to the livid colouring associated with such plants as Broom-rapes and Toothworts. When the colour is deeper an extensive and distinctive range of the so-called 'art colours' is produced such as soft pinks, rose, beige, café-au-lait, brown, blue, green-grey, etc. These quiet colours have an attraction of their own. One of the best of them is a pure silvery blue, the merits of which have led to the seeds being marketed under the description of 'Auricula Blue Shades'. This comes reasonably true to type and is well worth growing. A drift of it against a background of weather-worn Westmorland stone, with its grey-green foliage and flowers with pale sulphur-yellow centres setting off the exquisite blue of the corolla, makes a picture long to be retained in memory.

Although the group is overshadowed by the brilliancy of the Alpines, a representative set of varieties would make a great appeal to those appreciating quiet colour effects. To obtain it would require some effort because seed, apart from that of 'Blue Shades', is not to be found on the market. But plants are in abundance in cottage gardens, and a tray of spare Alpines for exchange and, maybe, a shilling or two can be counted on to secure good hunting. Then even with a small collection a little cross-breeding and selection should result in a group of plants unlike anything other Auricula growers could show.

The Striped or Flaked Auriculas are a group which has long since disappeared from cultivation. It was being grown in the earliest days of the plants' cultivation and, judging from the enthusiastic accounts of its growers, it was very much in favour with them. The following quotation from a manuscript dated 1732, published in the Annual Report for 1934 of the National Auricula and Primula Society (Southern Section), although written late in their history, agrees well with the brief descriptions of many sorts which were being grown a century earlier and forms a good introduction to the group:

These flakes, or striped Auriculas, have all their partisans, but they do not absolutely hold the first rank. Nevertheless they are highly valuable when they are glossy and look like velvet, when their stripes are clear and neatly divided from the bottom to the edge of the flower; the stripes are always either white or yellow; the whiter the stripes are, the finer; if yellow, the more like gold, the more agreeable. The bottom ought to be perfectly round, and not any ways angular, lest it

69

should so fall out, that the stripes, mixing with the bottom, would render the flower very disagreeable, if not insupportable . . . There are many of them larger, or wider, than a broad crown piece.

The comparison suggests that the flowers were particularly large but it is discounted to some extent by a further statement to the effect that 'the Pures (i.e. Alpines) are preferred to striped flakes and bizarres because they are generally larger and more resemble velvet'.

From this and a number of very similar, though briefer, descriptions it is clear that the striping was the result of the superficial colour-layer developing in strips instead of uniformly over the petals. The intervening areas then displayed the unaltered yellow or white ground colour usually only to be seen in the centres of the normal flower.

But one description of what its growers used to call this 'ennobled flower' shows that striping was a more complex phenomenon than would be expected from this explanation. In this it is unmistakably clear that some of the sorts had both yellow and white stripes. This raises the question of how both yellow and white can occur in a flower the background of which is either one or the other colour. The only way of throwing light on the problem seemed to be to find specimens of, or to rebuild, this lost group of plants. There seemed to be a good possibility of finding striped flowers in the extraordinarily heterogeneous mixture on sale on market stalls, or, failing this, it might still be carried in some few plants in a hidden condition as a recessive character (p. 146). The hunt produced no result for some years. Then when visiting an Alpine nursery a plant was found bearing faded flowers which suggested that it was a scarlet stripe. Although the poor condition of the plant made it seem certain that it would not survive much longer, gamble or no gamble, it had to be acquired. It flowered very early in the following season and the two miserable flowers it produced before it died a few weeks later had both yellow and white stripes. The yellow were those to be anticipated in a scarlet flower and the white were yellow stripes which had been masked by a thick coating of meal.

From the scanty pollen obtained from this plant a start was made towards raising more striped sorts (p. 147); then in the following year a plant showing some indications of striping was found in a batch of market plants and this also was brought into use as a parent. The striped flowers which have been obtained from these crosses belong to the Alpine and the Self groups. The Alpine set consist of only six plants at present with either scarlet and yellow or clear purple and white striping. As show flowers they have no merit but they have a gay inconsequence of their own; they are the ragamuffins of the Auricula world. The Self set of some twenty plants are a duller lot. They are purple or chocolate in colour with densely mealy, symmetrically placed, white stripes and less prominent yellow ones. They are interesting as curiosities but horticulturally of no value.

Our standards of taste have changed considerably with the passing of the years and nowadays striping is hardly tolerated in any flower except the carnation. It is consequently difficult for us to appreciate why the old-world growers were enthusiastic enough about the striped sorts to pay almost fabulous sums for new and distinctive kinds.

The 1732 manuscript from which the general description of the striped flowers was taken refers also to another group, the Bizarres, evidently closely connected with them. These are said to have been of two sorts,

... the old and the new ones. The old kinds are those which have a bottom of a different colour from the stripes; the most common have a white bottom with a yellow or gilt stripe; this kind is of a common size, and very apt to degenerate. The new kinds of Bizarrs are either raised in England, or from seeds brought from thence. They are admirably variegated, and charm by the great quantity of colours, which you often find as different, even in the same flower, as white from black. These are not generally so large as the others; but this fault may be corrected by sowing. They are commonly covered with a very fine powder, which being laid very thick on the bottom, renders them distinguishably brilliant.

Moreover they were very difficult to obtain. This

is an isolated statement and no further information about them has been found so far. So the mystery of this rather problematical group can be left for the future to solve.

The Doubles need no detailed description of their characteristics for the name provides all that is necessary. They form a group which is almost lost to cultivation, and nowadays the surviving examples are hardly to be found outside the collections of a few enthusiastic fanciers. So rare are they that most Auricula growers have never seen a specimen. They were never grown on an extensive scale, even between the years 1650 and 1750, from which most of their descriptions date. In fact the possession of a plant or two seems to have given their owners some of the prestige attached to those owning some acknowledged masterpiece of art. Now little more can be done than to read about them and try to picture the flowers once considered to be the most ennobled of all the Auriculas.

This is a tantalizing substitute for seeing the flowers themselves, for they were a greater departure from their single type than is the case with most flowers which have existed as double forms. The old doubles were vigorous, leafy plants with unusually large flowers, the petals of which were frequently marked with yellow or white stripes. If allowance is made for the paucity of the records it would seem that their colour range was a fairly wide one. It

included peach, crimson, crimson with yellow or
white stripes, purple, purple with white stripes,
black and liver-coloured with yellow stripes. The
sorts still in existence are also vigorous growers with,
probably, smaller flowers than their ancestors; and
their colours, reddish brown, brownish green, old
rose, and black, would tend to bar them from any
collections except those of the 'one of each' kind. The
only sort I have had lengthy experience of was a
purple brown given me some thirty years ago, and
all that its donor knew about it was that he had
grown it for many years. It was an interesting plant
for one never knew whether it was going to produce
trusses of double, double and single, or single flowers.
But it was a disappointing plant, for the hope that
pollen obtained from the single flowers might prove
valuable for raising new doubles turned out to be a
vain one.

Failure also attended an attempt to raise doubles
from an Alpine and a Green-edge, both of which
often produced flowers in which one or rarely two of
the stamens were petal-like. A flower with four
extra petals with little resemblance to the type one
had in mind was all there was to show from two
large batches of seedlings. The other doubles I have
had from time to time simply appeared and just as
promptly disappeared. Five of them were the first
trusses to develop on seedlings raised from crosses
between Green-edges. The only other one appeared

on a White-edge which for years had produced nothing but normal trusses. All of these flowers were fully double and green in colour and a single

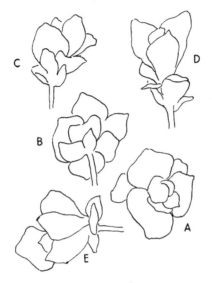

FIGURE 1

Double Auriculas: flowers consisting of green leaves only. A and B front and back, C, D, E, side views.

specimen nipped off and placed on the soil might easily have been mistaken for a growing rosette of a young plant (Fig. 1). The resemblance was a superficial one however, for the doubles failed to root and give independent plants. This White-edged plant happened to show other interesting features and so, in the hope of investigating these and obtaining more doubles, a stock of about a hundred plants was.

worked up from its freely produced offsets. There was not a double in the whole lot! The five plants from the Green-edge crosses too have never since thrown a double flower. So the gaily striped doubles of one's fancy are still only in cultivation in the land of dreams!

The doubling of flowers is a subject full of tempting problems for the investigator. In some flowers, such as the carnation, for instance, the problems are comparatively simple but in others, of which the Auricula is a good example, they are obviously of some complexity. The erratically flowering and the 'come and go' plants just described are almost certainly of the type botanists call chimaeras, that is to say plants in which the tissues are not uniform but a patchwork of various kinds. The isolation and propagation of any particular patch seems feasible enough but so far the right technique for the operation has not been found. The occurrence of a fair number of distinct sorts of double suggests that they may have arisen, not as sports but from seed, just as double-flowered stocks are raised from seeds of single-flowered plants. But how to set about finding this hypothetical plant is a problem still to be faced. For the present then the Auricula fancier who required doubles cannot expect help from the plant-breeder. At the most he can hope to experience the thrill of finding one amongst a batch of seedlings. The odds against such a find are tremendous but the optimist may remember the fact that the starting point of each group of Auriculas can only be described as some such lucky find.

In the Edged group the appearance of two new features not only sharply differentiates it from all of the preceding groups but results in the formation of a flower unlike that of any other plant. These features are the highly distinctive paste and the division of the corolla into two zones, the inner of which shows the usual Auricula colours whilst the outer is green or grey-green. The flowers too are more massive than those of the Alpine and Border groups. These new features so completely alter the appearance of the flower that anyone with only a rudimentary knowledge of horticulture might be forgiven for thinking that the Edged groups belonged to some other quite different kind of plant. In this group the richness of colouring which makes the Alpines so effective as border plants is wanting, and instead of the flamboyant effect of the Alpines the Edged flowers have a quiet refinement altogether their own. Their charm is indisputable and once anyone has fallen under their spell there is no other flower which can take their place.

The group is divided into three classes, distinguished from one another by the colouring of the outermost zone or edge. This is either green, grey or white. As the difference between these classes is mainly one of colour, the characteristics common to all three may be described first, leaving minor details for consideration later.

The paste, the most distinctive feature of the group, has already been broadly described (p. 17). The florist, from the very beginning, has attached much importance to its properties and no flower showing a faulty feature in any one of them was ever considered worthy of serious consideration. The first requirement is that the paste must be of a pure white colour. This is an easily satisfied one, for the only fault—a slightly grey tinge—is not of common occurrence. Next, the paste must have a smooth texture. This is often likened to that of the smooth surface of a piece of chalk but, somehow, this seems alien to its true character. It is difficult to describe but perhaps the matt surface seen when either white wheaten or potato flour is smoothed down with a knife blade will give a fairly accurate idea of the desirable texture. The smooth uniformity may be marred by a minute cracking which gives the surface a tesselated appearance, or it may be replaced by a coarse granulation not unlike that of the curd of a cauliflower. In outline the shape must be circular: the commonest departures from this are a scalloped or a polygonal edging, the latter being generally associated with angular shapes in the other zones of the flower.

The coloured zone, always known in this group as the body colour, ought to be marked off as sharply as possible from the paste on which it abuts; that is, its inner margin should be smoothly circular, but its outer margin is not definite, for it sprays out into the green zone in short irregular strips. This provides the one slight touch of irregularity in the formal up-building of the flower. Flowers in which the circular ring of the body colour is replaced by a perfectly hexagonal ring are frequently found in the course of Auricula breeding. The younger flowers of the truss then have pentagonal rings and the general effect is peculiarly irritating even though the flowers are sufficiently symmetrical.

The ideal shape of the edge of the flower is again a circle. In the Edged group, however, there is so marked a tendency for it to be polygonal or even star-shaped that small departures from the ideal have to be permitted. In this respect there is a pronounced difference between the Edged and the Alpine groups. The tube also differs in these groups. In the Alpines the colour of the interior is the same as that of the centre but in the Edged group it is some shade of yellow, and the purer it is the better, since the opening of the tube provides a small patch of colour in the heart of the flower. There is one exception to this, for the tube colour is white or greenish white in flowers in which the body colour is violet. This is often a useful indication of the true colour when it is so intense that it appears to be black. A cylindrical shape is desirable, since, if it

79

widens out and becomes trumpet-shaped, the middle of the flower becomes too prominent. Finally, flatness of the flower is most desirable; though again a shallow saucer-shape has to be allowed to pass muster. Waving, however, is intolerable, for it distorts the clear-cut circular outlines on which the beauty of the flowers is so dependent. This list of points with their corresponding faults is a formidable one, and it can safely be said that there is no other flower on which florists have made such extensive demands.

The name Green-edge defines this class sufficiently. The green is of various shades between the green of grass and that of the foliage of the yew. Theoretically it should extend to the extreme margin of the corolla, but whilst this is true of the majority of sorts an appreciable percentage of the modern varieties have small white patches of meal on their edges. If these are not conspicuous a judge will overlook their presence, but the old Lancashire florists had no hesitation about classing them as China-edged. The most prevalent body colour nowadays is black, the black being the deepest possible shade of such colours as crimson, purple, brown, etc. This tendency to avoid the more pronounced colourings appears to be on the increase, and it also seems that a wider green and consequently a narrower body colour zone is meeting with general approval. If the old convention that these zones should be about equal in width is departed from, then the move is in the right direction, for a broad band of dark colour produces a rather sombre effect. Another body colour, and one of singular beauty, is a shade of violet which harmonizes perfectly with the green of the edge. It has been a favourite since the early days of Edged Auriculas but even then growers complained of its comparative scarcity. Though it is no more abundant nowadays, a few kinds from which a choice can be made are available in the market. Varieties in which the body colour is crimson are also obtainable, and again the same story of their relative scarcity can be told. Plants of these crimson-flowered sorts scattered amongst the darker ones may make welcome spots of colour, but to many the clash of crimson and green is inconsistent with the quiet refinement associated with this class.

The scarcity of other body colours is surprising to anyone who has bred Auriculas at all extensively, and the lack of the rich golden browns, for instance, which blend so perfectly with the greens is regrettable. This raises the question whether, in the search for quiet perfection, extreme simplicity, relying mainly on formality and proportion, is not being overdone. The avoidance of inharmonious colours is clearly wise, but can this be said of the rejection of some of the subtlest harmonies to be found in any flower?

This account may suggest that a group of our

modern Green-edges with their simple colour scheme of green, black and white would have a monotonous appearance. This is true to some extent, but it is surprising what a diversity is produced by slight variations in the proportions and colouring of the different sorts. After all, it is the individual plant that can be picked up and carefully scrutinized which most interests the Auricula fancier. Where the class shows monotony is in its foliage, which throughout is a uniform grassy green very like that of the Alpines.

The essential difference between the Grey- and the Green-edged class is that the flowers of the former bear a coating of meal. The foliage may be either mealy or meal-less. In other respects the two classes are identical although strikingly different in appearance. The class is a much appreciated one, and quite a large number of varieties is obtainable from nurserymen specializing on Auriculas and probably a still larger number are in the hands of amateur breeders.

The most generally grown type of this class is one in which the Auricula has reached its finest development. In this the extreme edge of the flower is outlined with a distinct margin of silvery meal. Immediately within this is a zone of grey or greenish-grey colour encircling the, usually, deeply toned body colour. The silver, the green-grey, the white of the paste, and the body colour together produce one of the most exquisite colour schemes to be found amongst the Auriculas or indeed in any other plant. With such a complex of characteristics it is clear that many sorts of this type must exist, and although the general description is sufficient to identify the type it is inadequate for distinguishing between the numerous kinds now available. The task may be left to the expert. The rigid application of the fanciers' conventions has led to a standardising of the type; in its favour it may be said that the result is the survival of only the most perfect sorts even though some of them come under the category of 'too much alike'.

In another distinct type the meal is so sparsely developed that the individual glandular hairs are just distinguishable as minute silvery spangles against the green background. The colour effect is then a silvery grey-green, not unlike that of a willow leaf. Increasing quantities of meal and the conse-quent masking of the background add further shades of grey-green until finally the edge of the flowers becomes practically white.

Again, as with the Green-edges, those sorts in which the body colour is black are in most demand. But there are many other colours such as violet, crimson, rose, purple, brick red, Vandyke brown, cinnamon, etc. Whether this is a more extensive range than that of the Green-edges is uncertain, for no comparison is possible owing to the far larger number of Grey-edges that have been available for examination. A few varieties with one or the other of these colours were on the market in 1939 but whether they have survived the destruction of stocks resulting from the necessities of food production during the war is unknown.

Some of these colours associate particularly well with grey and grey-green, and suggest an interesting task for the fancier wishing to build up a distinctive collection of his own. But owing to the number of characteristics in this class which have to be taken into consideration this will be by no means easy.

The White-edged class is not grown quite so extensively as either of the two preceding ones. In fact some fanciers grudge it a place in their collections, saying that the flowers have too artificial an appearance. The briefness of the definition, White, is liable to cause some confusion between heavily mealed Grey-edges and the true White-edges, and it should be extended by adding that the meal should be of the fine type (p. 50). This forms dense incrustations which mask the tissues below very effectively, and although the meal may be less dense than in the paste, it nevertheless has a singularly pure white colour. The colour scheme is thus merely that of the body colour and white. The favoured body colour is again black, with the result that the flower has the simplest of colour schemes, if indeed it may be called one. The beauty of the flowers then is determined

84

in part by a simple, sharply defined pattern contrasting strongly with its background. But another feature is important. The main characteristics of the flower can be represented very accurately by drawing with indian ink on Bristol board but the representation altogether lacks the charm of the original. The difference between the two is that the drawing gives no hint of the texture of the flower, and this brings home in no uncertain fashion the outstanding part the meal plays in producing the distinctive appearance of both the White- and the Grey-edged classes.

The only other body colours available now are the deeply toned violet and purple seen in the Green- and Grey-edged classes. Others could probably be raised without difficulty but where sharpness of contrast is of such importance the paler tints are hardly required.

The foliage is similar to that of the Grey-edged class, but the meal, when present, is generally of the finer type. Where it is absent the pure white of the trusses arising from a bright green rosette has a somewhat incongruous effect.

The Fancies form a group which apparently has never been defined. The nearest approach to a definition found so far is 'any flower which does not belong to one or the other of the previous classes', the Fancies here being the last group mentioned. The group is thus a convenient pigeon-hole into which unsolved 'difficulties' can be consigned. Again nothing can be made of it from a study of the few descriptions existing of either the older or newer sorts. There are, however, a few kinds in cultivation which are characterized by the absence of body colour but which in other respects are Edged flowers. This is taken as the criterion of the group in the following account of it.

From the florists' standpoint the Fancies are not and never have been of much importance, but the theoretical interest of the group is sufficient to warrant a description of it. There are two types representative of it which are occasionally seen on the show bench. The first of these is green with the colour extending to the paste and usually with inconspicuous flecks of golden yellow at the points of junction of the petals. The other, often listed as a 'yellow ground Fancy', is more common, and a small number of fairly distinct sorts of it are in cultivation. This is a bright buttercup yellow in the region occupied by the body colour in the Edged group, and the tips of the petals, which may or may not be mealy, are green. Starting from these a large series of Fancies has been bred, and the types to be expected when the Edged sorts are robbed of their body colour have, for the most part, been secured. They include various shades of yellow and buff. The most interesting feature about them was that the yellow-grounded sorts could be arranged in a continuous series showing a gradual extension of the ground colour and a corresponding diminution of the green area. The green of the tips finally disappeared, leaving, where meal was present, only minute flecks. The end product of the series could only be described as a yellow Self. The complete sequence was only observed in the yellow-ground sorts and it did not extend to the production of a buff-ground Self.

The breeding of new Fancies is greatly simplified by their lack of body colour and some of the few sorts tested have been found to breed remarkably true to type. Horticulturally the group can have little value, for in only a few of the varieties is there any stability about the colouring. For instance the early flowers giving promise of an attractive green truss are too often succeeded by dully coloured buff flowers and even the trusses of brightly coloured yellow flowers may suffer from the same defect. But where this change does not occur the Fancies may have some decorative use as pot plants. One striking feature about the many plants under investigation was the general excellence of the paste. Its proportions were often bad, erring particularly on the narrow side, but the colour and texture were faultless.

The Selfs have flowers which are uniformly coloured from the paste to the margin. Thus they have some resemblance to the white-centred, non-gradated Alpines. There are good grounds for considering that they were in cultivation at the beginning of the nineteenth century but faultless descriptions of the early sorts are practically non-existent. Their historical study too is hopelessly befogged by the name Selfs being used for Alpines even as late as the twentieth century.

The rich colouring of the flowers which sets off the paste to perfection has made the group popular even with those florists who set little store by the Alpines. Its range is a particularly wide one and some of the colours such as scarlet, violet and blue seem to be purer than in any other Auriculas. The series begins with white—a dull isabelline white with its dinginess emphasized by the purity of that of the paste—and ends with a velvety black of almost unbelievable intensity, heightened again, if this is possible, by the brilliancy of the paste. This magpie of a flower attracts the attention even of those with no knowledge of Auriculas whenever it is exhibited. Between these extremes there is every colour which hirsutin is capable of producing.

Owing to the comparatively few features presented by the flowers, faults are not numerous. The most prevalent is an excessive narrowing or widening of the ring of paste, and in selecting plants the old 1 : 3 : 6 ratio (p. 22) should be kept in mind.

The vigorous foliage usually forming shapely rosettes may be either coated with meal or of a grassy green colour.

CHAPTER V

THE ORIGINS OF THE AURICULA

UNTIL the year 1875 the most complete accounts of the origin of the Auricula were those given by Gerarde in 1597 and Clusius in 1583 and 1601, in which they stated that it was a wild plant found growing in the Alps. In 1875, however, an Austrian botanist, Anton Kerner,[1] recorded a number of field observations which formed the starting point of an unexpectedly complete explanation of its real origin. Whilst collecting near Innsbruck he found specimens of the rare Primula described by Jacquin under the name of *P. pubescens*. Many botanists would have been content with making a note of its locality in their floras, but Kerner, who had an unusual knowledge of Alpine plants and also a special interest from his student days onwards in natural hybrids, speculated on the possibility of *P. pubescens* being a hybrid between *P. auricula* and *P. hirsuta*, both of which were flowering in the

[1] A. Kerner, 'Die Geschichte der Aurikel', *Zeitschr. des deutschen u. oesterreichischen Alpenverein*, VI, 39, 1875; 'Die Primulaceen-Bastarde der Alpen', *Oesterreich. bot. Zeitschr.*, XXV, 122, 1875; and 'Botanische Neuigkeiten aus der Gegend von Innsbruck', *Oesterreich. bot. Zeitschr.* XVII, 197, 1867.

immediate neighbourhood. The idea was far from being an obvious one, for the three plants, especially to a non-botanist, were manifestly totally different from one another. Kerner's examination, however, showed that *P. pubescens* had some features in common with the two other species; it had an eye like that of *P. hirsuta* but with the yellow colouring of *P. auricula;* and the general colouring of *P. pubescens*, a reddish brown or burnt sienna, was a blend of the yellow of *P. auricula* and rosy carmine of *P. hirsuta.*

Kerner concluded that the natural hybrid, *P. pubescens*, was the origin of the garden Auricula, now often botanically named *P. hortensis* Wettstein. The descriptions of the parent species, which Kerner thought had given rise by hybridization to *P. pubescens*, are as follows.

Primula auricula, L., one of the glories of the Alps, ranges eastwards from Dauphiné and Savoy through Switzerland and Austria to Serbia, and southwards into the Apennines. It is typically a limestone plant but specimens of it are occasionally found in peaty situations. It occurs in dense masses in pastures, on sunny cliffs, and especially luxuriantly in moraines. The flowers, generally scented and borne in trusses standing well above the foliage, are of a rich imperial yellow, the colour extending to the throat, which is surrounded by a thin powdering of meal. The foliage is usually heavily coated with meal. The leaves vary considerably in shape, the broadest being roughly obovate, the narrowest spatulate. The species is by no means a sharply defined one and several of the more distinct forms are generally recognised as sub-species. The most important of these, *P. a. Bauhini*, is easily distinguished from the others by its large rosettes and its heavily mealed leaves, each of which has the margin defined by a clear-cut silvery line. The trusses are many-flowered and the flowers are similar to those of the type except for the fact that the meal often forms a clean white ring round the throat. *P. a. ciliata* differs from *P. auricula* proper in having small, deeply notched leaves which are entirely free from meal. In addition to these there is a pure white-flowered variety of the type in cultivation. All of them are first-class plants for the alpine garden.

Primula hirsuta All.[1] is also a widely distributed species. It is common throughout Switzerland, the Austrian Tyrol, the Alps of Piedmont and Lombardy and in the Pyrenees, but usually only on granite or schist formations, where its general habitat is amongst rocks and stony pastures. The broadly obovate or rhomboidal leaves are of a grassy green

---

[1] The specific name *hirsuta*, after doing duty for some 150 years, is now to be changed to *rubra*, J. F. Gmel., according to the International Rules of Botanical Nomenclature (see W. Wright-Smith and H. R. Fletcher (1948), p. 677, in list of additional references at the end of the book).

91

colour and entirely free from meal. The short-stemmed trusses are rarely longer that the leaves and bear large flowers which are generally of a rosy carmine colour, although the colour may range from white through rose to a deep beetroot crimson. At the centre of the flower is a sharply defined white eye. The petals, which have a deep central notch, are either narrow or wide, with the result that the flowers are either starry or well rounded. The various forms of it are of no particular horticultural importance.

Both *P. auricula* and *P. hirsuta* have given rise, in their native habitats, to a number of hybrids with other species of Primula. Some of these have been brought into cultivation but they do not appear to have taken any part in the development of the Auricula.

Since Kerner's time, Wettstein (1920)[1] and Ernst and Moser (1926)[1] have shown that the artificial hybrid between *Primula auricula* and *P. hirsuta* is relatively fertile, in contrast to the sterility commonly associated with species hybrids. *P. pubescens*, by many considered to be the *Auricula ursi* II of Clusius, and the presumed natural hybrid between these species, was probably also relatively fertile; thus, following Kerner's views, the fertility of the garden Auricula, so valuable a feature from the florists' standpoint, can be accounted for.

[1] See additional references at the end of the book.

Historical investigation has shown that the first groups of garden Auriculas to come into cultivation were the Alpines (with meal-free leaves) and the Borders (with mealy leaves), and that they were probably derived from the Bear's ear (*Auricula ursi* II of Clusius) first grown, as far as is known, in the gardens of Vienna in the sixteenth century and thence distributed to other parts of Europe.

The colour phenomena of Alpine and Border Auriculas are complex, and their complete investigation would demand growing Auriculas on a scale which cannot be contemplated. It can be said, however, that the whole range lies within that anticipated from a knowledge of the properties of hirsutin (the pigment of *Primula hirsuta*) and the pigment of *P. auricula*, and that the various colours can all be found in these groups. Many shades, especially of red and orange-brown, no longer appear in cultivation, though, no doubt, they were represented in the collections of the seventeenth-century growers. One is worth an incidental mention, namely a green described on several occasions in the literature of this period. This owes its colour to a superimposed layer of lavender-blue on a yellow ground. Its colour thus has nothing to do with that of the Green-edges.

The above deductions provide an unexpectedly complete account of the origin of the Alpine and Border groups. They appear to be final as far as the former is concerned but more information is required with regard to the latter. The deductions also fit the facts as far as many Borders are concerned, but in this group there are many sorts in which the ground colour is buff instead of yellow. The buff is probably a derivative of the yellow colour brought in by *Primula auricula*, for several, normally yellow-flowered plants, have buff-coloured varieties. However, too little is known about the colour range of this species for this to be considered as more than a working hypothesis.

Kerner's views were the subject of a lengthy debate at the Royal Horticultural Society's Primula Conference in 1886. The result of this indicated a marked difference of opinion between florists and botanists. It opened with a mildly provocative paper by one of the former, S. Hibberd (1886), the gist of which is summed up in the following quotation: 'To obtain the two great classes of Auricula from *Primula pubescens* is a greater extravagance on the part of Professor Kerner than any florist has ventured on as yet'. The botanists, on the other hand, whilst accepting Kerner's views, considered that other Primula species must have played a part in contributing the blue shades, a view which, natural enough at the time, has been discounted by increased knowledge of the colouring matter of flowers.

One side of the *Primula pubescens* problem was omitted, strangely enough considering the botanists taking part in the discussions. This deals with nomenclature, for when Jacquin wrote a description of his newly discovered specimen and gave it a specific name he unwittingly caused an endless amount of trouble. It starts with the question of whether the immediate descendants of *P. pubescens* should bear its name even if the plants are almost identical with the original parents of this hybrid. From the systematists' point of view this is absurd, but gardeners still sometimes insist that *P. pubescens* is the correct name not only for these plants but for all known Auriculas. It is time the name was confined to plants rigidly conforming with Jacquin's original description of the plant which he wrongly, but excusably, considered to be a true species. The wide range of colour and the neat habit of growth of these primitive Auriculas make them attractive rock-garden plants. Recognizing this, Correvon of Geneva worked up a collection of them; but, although listed in his catalogues for several years, there was no great demand for them and nowadays specimens are hard to come by.

Kerner's views about *P. pubescens*, however, had no direct bearing on the origin of the Edged Auriculas. His main concern was with the characters occurring in the two wild, ancestral species, neither of which showed any sign of the outstanding features associated with the various Edged groups. Apparently, then, a new phenomenon had arisen, one of

95

some importance because it was clearly different from any that florists and botanists were familiar with.

The outstanding differences between the flowers of the Alpine and Edged Auriculas had inevitably led to many speculations on the nature of the green edging. The only conclusion reached was that the flowers of the latter group were unique. This certainly has a good deal of truth in it, and for the last half century or so the majority of the accounts dealing with Auriculas in the gardening press finish with a statement to this effect. There is an air of finality about the statement too, implying that the reason for the uniqueness of the Edged types was a mystery which was unlikely to be cleared up. Its solution, however, turned out to be a simple one, and once the true nature of the green edging was realized the distinctive features of the edged flowers found a ready explanation. This in turn led to the tracing of the origin of each of the groups distinguished from the more primitive Alpines by the presence of a paste.

The starting point came after an examination of the minute structure of the green edge. The details of this need find no place here, and it is only necessary to say that, in every respect, the structure was found to be precisely that of a foliage leaf. The identity was so complete that sections of the two when mixed together were indistinguishable. In

96

this type of flower, then, the petals have been replaced by leafy structures. In the homely language of the gardener it would be said to have 'thrown a green sport' whilst botanists would describe it as having become 'virescent'. The word 'sport' is now used to cover a number of unrelated phenomena, and for the sake of clarity it will be abandoned here and the word 'mutation' used in its place. The distinguishing characteristics of a mutation are that it is a feature hitherto unknown in the plant, that it is hereditary, and that when crossed with the normal it generally behaves as a recessive character (p. 146).

Very little is known at present about the phenomenon of Virescence. This is due in part to a neglect to investigate what is obviously an abnormality and in part to the rarity of its occurrence. It is so infrequent that it would be fairly safe to say that the examination of a million plants would not result in finding a single virescent specimen. By way of example the story of a lengthy hunt for them in two other members of the Primulaceae may be told. The virescent form of the common primrose is of particular importance, for, as the sequel will show, there is a noteworthy parallelism between its development and that of the Auricula. The story started over half a century ago with the finding, whilst scrambling through the undergrowth of a wood full of primroses, of a plant in which the yellow petals had been replaced by small but unquestionable leaves. Since then a look-out has been kept every spring for other specimens, but although an almost astronomical number of primroses has been looked over I have seen no other virescent specimen growing in the wild state. Another find, vouched for by the receipt of flowers matching those of the plant then being grown in the garden, was made some twenty years later, a third well authenticated specimen was reported from Argyllshire in 1924, and three years ago yet another was reported in 'The Times' with a description which could leave no doubt as to its genuineness. A still less successful hunt has been made through what amounts to acres of the bird's-eye primrose (*P. farinosa*), the only other species available in sufficient quantity to give any hope of a lucky find.

Virescence has not been recorded in any of the floras dealing with either of the parent species of the Auricula, but this cannot be taken as evidence that it never occurs; for even if botanists, whose business it is to draw up meticulously accurate descriptions of the types, had found specimens, they would naturally leave such obvious abnormalities out of account. Virescence may be a more common phenomenon amongst cultivated plants; but here again the likelihood of its being reported is slight, since the majority of gardeners would have no hesitation in throwing out a plant which had merely 'gone green'. Here and there brief notes have been published on

98

green flowers of *Primula japonica* and *P. malacoides*. In the case of the primrose, however, this mutation was preserved, and as early as the middle of the sixteenth-century green primroses were well known garden plants and they had already given rise to double sorts and to single and double green poly-anthuses.

Why the analogy between virescent Auriculas and primroses was so long in being recognised is a problem. In the primrose the petals are replaced by leaves matching, with their characteristically corru-gated surfaces, those of the foliage so completely that the question of their nature can hardly arise. In the Auricula, however, the presence of the paste and the patch of body colour may have tended to obscure the likeness. But, as most of those who have raised large numbers of seedlings of the Edged Auriculas must know, corollas built up of five or six small leaves are met with frequently. It is here that the limitations imposed on the fancier become troublesome. To him an Auricula is not an Auricula if it has a star-shaped flower and is consequently unworthy of further notice.

Once the leafy nature of the corolla of the edged flowers is recognized it becomes possible to account for most of the peculiarities which place them in a group of their own. The old problem of why a flower which normally has a well rounded shape should so often have a polygonal or even a starry outline finds an explanation in the fact that the slightly notched, heart-shaped petals of the original type have been replaced by pointed leaves. This suggests that it might be worth while to select for breeding purposes those plants in which the foliage leaves have a rounded rather than a sharply pointed outline. If there is any correlation between the shapes of the flower and the leaf—and this has still to be proved—then it should be possible to build up a strain in which this troublesome feature was eliminated once and for all.

Shape, however, was not the only feature brought into the flower when leaves replaced the petals. Of greater importance, because it is a more desirable feature, is the fact that the leaves brought mealiness with them and so gave the flower much of the great diversity of character which is so notable in the foliage of the Auricula. The distribution of the meal on the foliage already described (p. 51), is matched by that on the corolla edges. In the Green-edged group, in which there is no sign of meal, the leaf type of the parent, *Primula hirsuta*, is represented. The silver-margined but otherwise meal-less green leaf finds its counterpart here in one uncommon sort in which the bright green corolla is delicately edged with silver. The classification of this sort may cause trouble to some harassed judge but the fact is clear enough. The Grey-edged, White-edged and Fancy groups all have more or less meal on their flowers which was brought in originally on the foliage of *Primula auricula*. In this parent the leaves are slightly, moderately, or extremely mealy, and again there may be a concentration of meal on the leaf margin. All of these types can be found in the flowers of even a small collection of the Grey-edges, whilst the fine type of meal has given rise to the White-edged group. Incidentally it was this mealiness which enabled the Auricula to outstrip its early rival, the primrose, though the latter had a start of at least two centuries.

So far this account has only referred to the green edging having a leaf-like nature, without throwing any light on any of the other problems associated with virescence. But the story can be carried further by a study of the calyx of the flower. This is the one part to which the florist has given no attention, doubtless because it was out of sight, and so questions of points and faults needed no consideration. This has an advantage, for it can be assumed that the calyx is in its natural condition and unaffected by any conscious attempt to alter its character. Observations on it are best made when the corollas have fallen. Then the leaf-like nature, as in the calyces of so many flowers, becomes clear enough. In the parents (*P. auricula* and *P. hirsuta*) the calyx is a small cup-like structure formed by the fusion of five sepals in such a manner that their upper halves are left free. The free portions are too small to be

101

clearly leaf-shaped. But in a collection of modern Auriculas the sepals are far larger and their leafiness cannot be questioned. Moreover they show all of

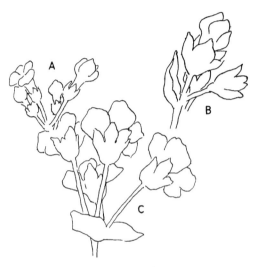

FIGURE 2

Calyces. A, *Primula auricula* with small tooth-like sepals; B an Alpine, and C a Green-edge, with large leaf-like sepals.

the diverse characteristics of the true foliage even down to differences in the leaf shapes. Their larger size is no doubt correlated with the increase in the size of the corolla which breeders, throughout the centuries, have been aiming at. The size, however, is very variable and in some flowers the diameter of the calyx falls little short of that of the corolla (Figs. 2 and 3).

Surrounding the ovary, at the base of the calyx, there is in most Auricula flowers a dense ring of fine meal, the only exceptions being found in some of the

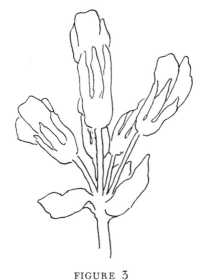

FIGURE 3

Enlarged calyx with narrow, spatulate sepals.

plants belonging to the meal-free Alpine group. The innermost edge of the ring is always sharply defined but its outer edge, more often than not, is somewhat diffuse. The result of this is that a truss of these expanded calyces has a general resemblance to a truss of an edged flower (Fig. 4). There is no ascertainable difference between this ring of fine meal and the paste of the corolla. Such a series of similarities cannot be fortuitous, and it is a fair inference to

103

state that the virescent condition of the edged flower was effected by the total replacement of the petals by sepals. This then provides, for the first time, an

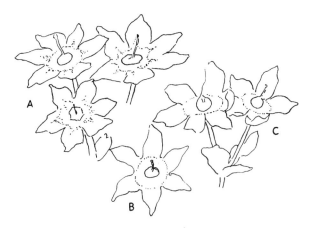

FIGURE 4

Large leafy calyces showing paste. A, Ill-defined, irregular paste. B and C, Paste with regular, circular outline.

explanation of how the Auricula's most distinctive feature, the paste, came into existence.

It is too much to hope that we shall ever know precisely what the mutation looked like which started the Auricula off on its extraordinary course of development. There is, however, a possibility, if it does not yet amount to a probability, that this mutation, or something very similar to it, is still appearing amongst the progeny of our present-day plants. This is a plant with green foliage a little under the average in size; green-edged flowers built up of broad, sharply pointed leaves; a mere trace of body colour on the inside but more on the outside of the flower; and a good paste except for the fact that its margin is diffuse. The flowers have a singular lasting capacity and generally only begin

FIGURE 5

Corolla of green 'leaves': practically no body colour, irregular outline to paste.

to fade in August (Fig. 5). Their stamens never develop fully and the ovary may or may not be normal. In the latter case the style is slender and flattened and the stigma is wanting, or again a group of three or sometimes four occur on the one ovary. The lack of pollen makes it impossible to self-pollinate the flower and so simplify the task of

arriving at some knowledge of its genetic constitution. What little information is available has resulted
from crosses made with two other Green-edges,
chosen as parents because something relevant was
known from previous experience with them. It was
known, for instance, that one of them carried, as a
latent recessive, the factor for meal production. In
the second generation the presumed mutant re-
appeared, together with Green- and Grey-edged
flowers with a considerable colour range; Selfs;
Fancies; a Double; and a monstrous ten-petalled
completely sterile flower which just failed to reach
a diameter of two inches. One noteworthy feature
of this complex mixture of sorts was the excellence
of the paste and the shape in a majority of the
flowers. This was hardly to be expected, for although
the two Green-edged parents were good florists'
flowers, the faults of the mutant were many. The
number of plants of these diverse sorts was too small
to have any statistical significance and only the
selfing of a few selected plants, if ever conditions
permit their cultivation on a sufficiently large scale,
can push the story further.

The Fancies, as defined in the previous chapter,
provide an example of the commonest of all mutations, namely the loss of colour. Such a loss is
generally the first to occur when a new species of
plant is brought into cultivation, and the frequency
of its occurrence is evident from the large number

106

of white varieties amongst garden plants. The flowers of the Fancies, however, are not white, for when the hirsutin pigment disappeared from a flower with coloured edges the green of the chlorophyll or the yellow of its derivative was left unchanged. Only one example of this mutation has been seen amongst all of the Auriculas kept under observation. The plant which provided it was a White-edge which when first noticed appeared to have a green thrum. In fact, the 'Thrum' consisted of some twenty or thirty small seedling-like plants which had taken the place of the seeds. Unfortunately it was assumed that this was a case of vivipary and that the small plants when grown up would be identical with their parent. Only a half dozen were picked off instead of all that were available. Four of these reached the flowering stage: three were precise replicas of the parent, the other was identical with it in every respect except for the fact that the deep body colour was replaced by a ring of buttercup yellow. It was thus a white-edged Fancy, and by some strange chance the one sort needed to complete the series (p. 86). The parent plant was one which had given rise to a double form, and a stock was worked up in the hopes of a repetition of this phenomenon, but, like the double, the 'viviparous' form failed to reappear.

The Selfs are evidently closely connected with the Edged groups for they have the distinctive paste in

107

common. Moreover, as all who have raised many plants of the latter group recognize, flowers with an especially wide zone of body colour may have their green or grey margins diminished to such an extent that it is only when the flower is held horizontally and viewed from the side that the existence of the margin is evident. The flower then is practically a Self. Undoubted Selfs, too, are consistently produced from the seed of self-pollinated Grey-edges. The commonest colour is a dull vinous purple or vinous mauve, and it is worth noting that seedlings from the plants so coloured come remarkably true to type. The frequency of this colour is due to the fact that the most approved type of a Grey-edge has a black body colour. If this is violet, then violet Selfs, often of great merit, occur. Similarly Grey-edges with a crimson body colour give rise to vivid crimson Selfs, and there seems no reason to doubt that a comprehensive testing of the many coloured sorts which still exist would account for most of the colours known amongst the Selfs. The one difficulty, however, is to explain the origin of the various yellow shades found in the group. These, in a sense, are colourless, since they lack the hirsutin pigment. In this respect they come close to the Fancies. The continuous transition series from green to yellow already described (p. 87) may then be an indication that the yellow Selfs originated from this group. But more investigation of this problem is necessary.

The origins of the groups characterized primarily by the possession of a paste now seem to be reasonably well established, and it may be assumed that the first group to come into existence was that of the Edged flowers, and that these in turn were the starting points of the Fancies and the Selfs. These two latter are thus comparatively recent developments and consequently some expectation of finding the actual date of their appearance is justified. Definite historical evidence is, however, wanting, thanks to the chaotic state of the classifications current during the nineteenth century.

In addition to the small group of mutations which determined the development of the Show Auriculas, a number of minor ones are known which, though of little interest to the florist, have some theoretical importance. One of these provides a starting point for Hose-in-hose Auriculas, corresponding to the old-world Hose-in-hose primrose. Such are by no means a rarity, and signs of them can generally be found in any moderately sized collection. These take the form of small tightly rolled structures, too inconspicuous to be noticed unless specially looked for, situated on the exterior of the upper half of the tube. Where larger numbers of plants are available it is evident that these structures are petaloid, for each one has the colouring and marking of the corolla above it. Those figured are the largest seen so far, but they clearly make a close approach to the

109

the supplementary corollas of the Hose-in-hose primrose (Fig. 6). Another mutation is akin to the equally
ancient primrose known as Jack-in-the-green. In

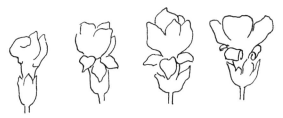

FIGURE 6

Beginnings of Hose-in-hose type of Auricula.

this, leafy sepals are developed to such an extent
that they overlap the petals, with the result that the
corolla seems to nestle inside a green jacket. Specimens of this have been received from time to time,
and it is certain that this departure from the normal
is not particularly rare. A still more eccentric form
in which the sepals have become full-sized foliage
leaves is illustrated in Fig. 7. It is probably a
chimaera, but there has been no opportunity yet of
determining whether it is hereditary as the Hose-in-
hose and the Jack-in-the-green mutations are.

The striking parallelism between the mutations of
the Auricula and primrose make it possible to hazard
a guess as to what other mutations may be looked for.
If for instance mutation 'Z' is known in the primrose,
then, by way of a working hypothesis, it may be
expected to occur in the Auricula. It is unlikely,

110

however, that much more progress in investigating mutations can be made on these hit-or-miss lines. There are undoubted chances of making rapid pro-

FIGURE 7

Monstrous flower with sepals almost the size of the foliage leaves, corolla normal.

gress on other lines in this direction, because plants can be induced to mutate by exposing them, at an early stage of flowering, to an appropriate dosage of X-rays. Amongst the plants raised from flowers so treated there is generally a considerable proportion of mutants. Whether any of them would have horticultural value remains to be seen for it is possible that all of them have already been found and utilized in the building up of the various groups

111

we now have. Anyhow whatever happens and however deeply these matters may be probed into in the future, the finding and propagation of the first virescent mutation will still remain a thing to wonder at.

CHAPTER VI

CULTIVATION

IT is natural to assume that when a plant has been in cultivation for over 300 years all of the details associated with growing it successfully are fully known. The experienced grower will probably say that this is so and that Auriculas are as easy to grow as Aspidistras. But that it is not so is evident from the fact that some who start off enthusiastically with a collection of purchased plants lose many of them within a year or so of the first repotting. Then they too often conclude that 'Auriculas dislike me' and give up the attempt to grow them; or it may be that, still intrigued by their beauty, they go on year by year replacing their losses by new purchases. Unfortunately stocks of Auriculas are very low at present and likely to remain so for some years owing to the slowness of their propagation. Nurserymen may then complain that they are wasting their time in growing plants destined to die in two or three years, and there is much to be said for this from their point of view.

But there is no reason why the grower should despair, for the trouble can almost invariably be traced to some easily corrected fault in cultivation which, once attended to, makes Auricula growing as simple as that of any other pot-grown plant. On the face of it diagnosing the fault is not easy, for satisfactory growth is dependent on a number of factors and the faulty one does not necessarily betray itself by giving rise to any particular symptoms. But in practice it can generally be assumed that the compost fails in some way to meet the special requirements of the Auricula. General experience and some preliminary experimental work indicate that the lack of really efficient drainage and consequently of soil aeration is the commonest of the possible causes of failure. Once a radical change in the constitution of the compost has been effected the novice can confidently look forward to better results.

There is little information available about Auricula cultivation in the early days. One of the best descriptions is that given by Bradley in *New Improvements of Planting and Gardening*, 1717. It is a straightforward account of the growing of the group now known as Alpines. But these were almost invariably grown as border plants and, though special selections might be potted up, the day for general pot cultivation had not arrived. It was not until a century later that a complete account of Auricula growing was published by I. Emmerton in 1815, *A Plain and Practical Treatise on the Culture and Management of the Auricula.* Round about the same period books on the same subject were published by Maddock and by Hogg, and for the first time in its history the Auricula had a literature of its own. Since then most of the literature concerning Auriculas is to be found in the more ephemeral publications of the gardening press.

Of these Emmerton's book is the most important, for he not only knew his subject thoroughly but he had the makings of a born experimenter. His thoroughness, too, was remarkable and no detail, even down to the pace at which his carriage should be driven when carrying plants for exhibition, was omitted. Except for a few minor discrepancies, chiefly with regard to the dates at which some operations should be carried out, the three accounts (Emmerton's, Maddock's and Hogg's) agree well with one another, and, moreover, conform to present-day practice except perhaps for the fact that they were written for growers who did not have at their disposal the varied assortment of artificial manures modern growers are tempted to make use of. All three writers were nurserymen accustomed to growing their plants under the best of conditions, and consequently they have little to say about the difficulties the beginner, even if a skilled gardener, may meet with. Their sections dealing with the preparation of composts are of no particular interest nowadays, but Emmerton's deserves a detailed

description partly because of its extraordinary nature
and partly because it is for ever being quoted in a
thoroughly garbled form.

He was convinced, if not obsessed by the idea, that
Auriculas required the richest possible diet, and he
set out to provide this by means of a compost as
extraordinary as any that has ever been devised. It
consisted of three parts of goose dung steeped in
bullock's blood, three of sugar-baker's scum, and two
of fine yellow loam to which a small quantity of sand
might be added if considered desirable. The sugar-
baker's scum, he understood, consisted of 'the dross
of sugar, a portion of the West India mould, fine
clay, bullock's blood, lime water, etc.' The dung and
blood were put into a pit covered with a hurdle to
keep dogs off, and left to putrefy—much to the
annoyance of his neighbours. In six weeks or so the
pasty mass solidified and was then ready for digging
out and mixing with the sugar-baker's scum and
loam. The mixture was exposed to the sun and air
for two years and the heap broken down and raked
through and through at monthly intervals. On its
richness, mellowness, freedom from insects, weed
seeds, and any other nauseous properties, Emmerton
insists time after time, wearying the reader of his
book with continuous repetition. That the compost
had some merits is proved by the fact that he was
a very successful exhibitor and that more than one
nurseryman of the period thought it worth while

to undertake the tedious and unpleasant work involved in its preparation.

Nowadays, however, the one thing that growers and writers seem to know about Emmerton is that he used bullock's blood in preparing his composts and all too often they modify this perfectly correct statement by altering it to 'in growing his Auriculas', leaving the impression that his cultivation was of a peculiarly macabre nature. Strangely enough the wheel has come full circle again and now dried blood is one of the feeding materials growers use extensively; incidentally it is far nearer the natural product than ever Emmerton's mellowed compost was. What reasons he had for devising this strange mixture can never be known, and it can only be looked upon as an example of the lengths to which an enthusiastic grower was capable of going in his efforts to outshine his competitors.

At this period Auriculas were grown very intensively and the tradition became established that they were gross feeders. This has died hard and it is only within the last generation that it has become generally realized that the plant can be grown to perfection on a moderately austere diet. The present-day composts usually consist of loam, sand and either leaf mould or some substitute for it. Dozens of specifications for them are to be found in the gardening papers, and no doubt those who have contributed them have found them valuable.

At the best they should be treated as guides only, for the grower has to learn by experience how to make the most of the materials he has at his disposal. These are far from being standardised, and two of them are about as variable as any inert substance can be. The worst in this respect is loam. This is stated by the dictionaries to be a mixture of clay and sand, and one can amplify the statement by calling the mixture in which sand preponderates a light loam whilst an excess of clay produces a heavy loam. To the gardener, however, loam generally means the top spit of an old pasture after the foliage and roots have more or less decomposed. But if this is cut from a pasture on a clay soil derived from such formations as the lias, boulder clay, gault, etc., it is little more than a pure clay. When thoroughly weathered by frost it may be porous enough to appear suitable for the preparation of potting soil, but if used for the purpose the clay particles coalesce again after a few waterings and form a pasty mass almost impervious to air and water.

It is a sound tradition that the Auricula needs good drainage for its proper development. This not only means that the pots must be crocked so that any free water will drain away immediately but also that the soil will drain itself and never become waterlogged. Auriculas growing in soil incapable of draining itself freely show fairly characteristic symptoms: the foliage is often paler than normal and it may be spotted with still paler or yellowish areas; few, if any, offsets are produced; and the root system is stunted, the main fibrous roots generally failing to extend to the side of the pot, with the more slender feeding roots forming a mat round them. Strong-growing varieties will struggle along in this unhappy condition for several years but the weaker ones may die after enduring it for a single season.

Sand is used primarily with the object of ensuring drainage through such impervious soils. But to make up a compost of the correct degree of porosity is by no means a simple task. It can only be done, if in some cases at all, by guessing at the quantity required, varying the amount both upwards and downwards, and then testing the mixtures by filling them into pots and heavily watering them. If the water runs through freely all is probably well, but even then the soil should be examined a day later to see that there are no signs of water-logging. A heavy loam may require the addition of an equal volume of sand, or even more, to make it sufficiently porous; a good average loam, similar in texture to the well known Kettering loam, about one part in three; whilst lighter sandy loams will require no further addition. Silver sand is the kind in most general use, and if a choice is possible the coarsest grained sort should be obtained. The kind often described as builder's sand is open to suspicion, for much of it is not

119

particularly porous and it may even set hard after continuous watering.

When a thoroughly free draining mixture has been obtained the next step is the apparently paradoxical one of modifying it so that it will retain a sufficiency of water and not dry out too rapidly. Besides leaf mould various substances such as dried-out cow or horse manure, spent hops and peat are used for this purpose, and at one time the use of 'punk' or decayed wood was advocated. The nutritive value of these materials is more or less negligible, their main value being that they soak up and retain moisture. Apart from finely ground Sorbex peat they are very variable in their composition. The value of leaf mould cannot be assessed by eye. It depends largely on the conditions obtaining during the decomposition of the leaves: a rich dark sample may be so acid as to be injurious whilst a light brown sample, even if threaded through by fungus mycelium, may be ideal for compost making. This variability is probably the reason for the growing practice of using peat in place of leaf mould. But here again a warning is necessary; for though it is amply proved that Sorbex is an excellent substitute it by no means follows that this is true of all of the sorts of peat made use of in gardening practice. No definite proportion of peat to the loam-sand mixture has been established but one in four or one in five is a commonly used ratio.

A compost with a fertile loam as its basis does not require any reinforcement by the addition of artificial manures. If, however, there is any doubt about its fertility a small quantity can be added of some slow-acting manure such as bone-meal. The amount necessary can, again, only be guessed but the addition of a 60-sized pot-full to a bushel of the mixture can do no harm and will probably have beneficial results.

Growers in large towns, where incidentally Auriculas often thrive well, may have some difficulty in obtaining the ingredients for compost making and those who have to be content with growing only a few dozen plants may not want to be bothered with making it up and storing it. They are then driven to purchase potting soil from a horticultural merchant or a nurseryman. When doing so any of the vendor's comments on its fertility may be ignored and attention concentrated on its texture. The soil will probably be in a bone-dry condition and an inspection of it will give little idea of its properties when moist, so a sample should be tested in order to make certain that, even if heavily watered, it shows no sign of water-logging. The quantity required can be estimated on the basis that one bushel is sufficient for about 50 pots—40 being 48s and 10, 60s.

With all these variables to deal with it is certain that elaborate specifications for the preparation of the compost are of no great practical value. The possibility, too, of mixing composts of unlike properties at various times is evident. This can be avoided by making notes on their constitution, and when one has been found to be satisfactory it is well to adhere to the specification year by year.

Growing Auriculas in the open instead of in pots raises a crop of problems which still require investigation. The Alpines and Borders require little consideration, for they can be grown on practically any kind of soil. The Show Auriculas are, of course, almost always grown in pots, but where large numbers of plants have to be dealt with, a shortage of glass may lead to their being grown in the open. The results can be very satisfactory even in soils which an expert grower would consider altogether unsuitable. It has been difficult to obtain any information which will help to account for this, and at present only the facts observed whilst growing plants under these conditions can be given. The soil concerned is a heavy loam overlying a clay which geological experts have described as either gault or boulder clay. After the winter's frost its tilth is excellent but unless the surface is cultivated at every opportunity the tilth is soon lost and the soil sets, becomes so rock-like that, when in a dry condition, it is difficult to penetrate with a fork. The drainage is consequently bad but nevertheless Show Auriculas thrive on it. Some measure of the suitability of the soil for their growth is provided by the fact that a few Green-edges put out 40 years ago, when the garden was first made, are still growing as vigorously as could be wished for.

This contradicts much of what has been stated above, regarding composts. Several spasmodic attempts have been made to clear the matter up. These took the form of growing plants in a series of composts based on this heavy loam. In the first trial it was mixed with an equal volume of silver sand, in the next two the proportion of sand to loam was as 2 : 1 and 4 : 1 respectively, and finally a moraine mixture of sand and chips containing only one part of loam to six of the drainage material was used. None of these mixtures broke down the plasticity of this clayey loam and after two seasons at the most the plants showed the unmistakable signs of water-logging. No sweeping conclusions can be drawn from so limited a set of trials, but they suggest that any loam in which the clay percentage is at all high is unsuitable for compost making. Just what is a safe percentage is not known, and rather than make a troublesome investigation to determine it, the difficulty can be avoided by obtaining a loam with a texture similar to that of Kettering loam.

These rather casual soil experiments led to one which had as its object the determination of whether Auriculas could be grown under the standardized conditions provided by sand cultures. In these, soil is replaced by sand and the necessary food materials are supplied when watering the plants. As far as growth is concerned the results are good but until the plants have flowered nothing further can be said about the value of the new method.

Auricula seed is notoriously tricky in its germination. The actual germinating capacity of home-grown seed or of seed from some reliable source is generally high, being of the order of 85 to 95 %, but the time of germination is erratic. Imported seed, particularly that of the parent species, frequently has an almost negligible germinating capacity. When the seed is sown immediately after it has been gathered one of three things may happen: it may germinate uniformly in about three weeks, every seed apparently growing; germination may be partial with only a few seedlings putting in an appearance instead of the abundant crop anticipated; or no germination occurs. But the seeds which fail to sprout in the early autumn may be expected with considerable certainty to do so early in the spring, even if the pans in which they have been sown have been kept dry throughout the winter. This delayed germination is not uncommon, especially with Alpine plants, and it can generally be accounted for by the failure of the essential water supply to penetrate the seed coat. But the usual methods of overcoming this difficulty by scratching the seeds with some abrasive, such as glass-paper, or by treating them with sulphuric acid has no effect on Auricula seeds, and it can only be concluded that the last stage of the ripening process is often carried on when the seeds appear to be in a completely dormant condition. As a consequence of this it is usually recommended that the seed should be sown in the early spring. But if germination can be secured in early autumn the young plants grow strongly until wintry conditions check growth and quite probably flower in the following spring. This saving of a year is often an important consideration and it pays to make the most of both seasons by sowing in the autumn and stacking any pans in which no germination has occurred under the staging until spring, when they are brought out and growth started by watering.

Seed should be sown as thinly as possible, and as there is not often a superabundance of it, mistakes in this direction are unlikely to occur. The compost may be either the one in general use or it may be diluted by the addition of more sand. Spacing the seeds rather than dribbling a pinch of them over the surface of the soil takes time but is well worth the trouble. The best plan is to mark lines, at right angles to each other, about one and a quarter inches apart on the surface of a pan of compost. Then at each intersection of the lines a single seed is placed. This can be done most easily if the seeds are spread out and picked up one by one on the pointed tip of a moistened glass rod. The rows are then covered with finely screened compost or sand, using a sufficient quantity to bury the seeds to a depth of an eighth of an inch. The first watering and any waterings required before the seedlings have a good root-hold must be made by placing the pan in water, the level of which should be about the same as that of the soil surface: watering with a can with even the finest of roses is too likely to disturb the seeds. The pan should then be covered with a sheet of glass and placed in fairly light surroundings. In the early months of the year the sunlight is not intense enough to damage the seedlings, nor again is the dullness of the light likely to cause them to become drawn. When the seedlings have four or six leaves and are large enough to handle without risk of damage, those in alternate rows should be lifted and set out in pans at intervals of two and a half inches; after this, plants from the remaining rows are transplanted in a similar manner. This spacing is sufficiently wide for flowering the plants in their first season of growth although they will become somewhat crowded. Alternatively, the seedlings can be pricked out round the edges of pots at one inch intervals and transferred, when they touch one another, to 60s for flowering. The former method has two advantages: the plants make better growth than they do in pots and the labour of watering is appreciably reduced.

When flowering begins the plants require screening from too intense sunlight. For this purpose controllable roller blinds are ideal provided there is always someone available to adjust them when advisable. A simpler and quite satisfactory method is to make use of slatted screens built of laths an inch wide spaced at one-inch intervals. These can remain *in situ* until the sunlight loses some of its intensity in the early autumn.

With the flowering of the plants the most interesting of the grower's tasks begins, namely, deciding which of them is worth preserving. By the time he has watched the buds slowly expand into fully open flowers he will have made up his mind on this point and be ready to label the selected plants. Wooden tallies will do for the pot plants, but, owing to the crowded conditions, they are less satisfactory for the plants in pans. In their place a six-inch piece of stiff wire, with a small piece of Whatman's or similar paper threaded for recording the number, and pushed into the soil as close as possible to the plant, will save any mistake at transplanting time.

Potting can be carried out at any time whilst the plants are growing but the best time for it is as soon after flowering as possible. At this time a fresh cycle of growth is beginning and new roots are pushing out from the buried portion of the stem. If the plant is carefully handled it can be placed in its new pot without causing damage, and the check to growth, inevitable with the disturbance, is reduced to the minimum.

The young plants which have flowered in 60s and those in pans require to be moved on to 48s. Some growers insist that the pots should be of the deep type known as 'Long Toms', which were almost universally used in the early days of Auricula growing; but the shape of the pot seems to be a matter of no importance.

There is some difference of opinion about the re-potting of older plants. It used to be the practice to do this annually, but if the plants are obviously growing strongly, re-potting in alternate years is now considered to be all that is necessary. Probably the growing use of artificial manures has a direct bearing on this point. With these older plants an increase in the size of the pot is unnecessary except, perhaps, when some unusually large specimens have to be dealt with. The object of re-potting is to replace with fresh material the partially exhausted compost, more especially its humus component, which decomposes in the course of a couple of seasons.

After pushing the plant out of its pot or lifting it from the pan the soil has to be removed from the root system by gentle pressure and by shaking. If root aphid is found to be present it can now be killed by washing the roots in soap and water or in a dilute solution of nicotine. Any roots which seem to be no longer functioning should be removed, and the 'carrot' should be cut back to the quick if its end is showing any signs of decay. Offsets too should be removed and set aside if required for propagation, and at the same time any small buds on the stem should be rubbed out. The object of this is to keep the plant to a single crown which will throw one fine truss rather than the two or three small trusses of a many-crowned plant. These operations involve some wounding of the tissues, and it is customary to dress any freshly exposed surfaces with powdered charcoal or flowers of sulphur. As an alternative, though not an attractive one, the cut surfaces may be allowed to dry and harden in the sun. After this cleaning-up process the plant is held centrally in a previously crocked pot and dry compost poured round the roots, firming it from time to time by sharply tapping the pot on the potting bench. Hard packing is inadvisable and nothing more than a slight firming of the soil round the collar is all that is necessary. When filling the pot care must be taken to leave a sufficiently large space between the soil surface and the rim of the pot to hold enough water to soak the soil thoroughly. A driblet of water sufficient only to moisten the top inch or so is useless to a developing plant which is pushing its roots towards the bottom of the pot. The plants, especially at flowering time, must have a steady supply of water, for the transpiration rate is high owing to the higher temperatures

129

under glass and the air stream through the ventilators.

The re-potted plants, after the dry compost has
been thoroughly soaked, should be placed in a well
shaded, draught-free position until they show signs
of recovery. In the course of a fortnight they should
have taken hold of the soil and started into fresh
growth. At this stage they are ready for transferring
to their summer quarters. Whilst in these every
effort has to be made to build up sturdy plants, for
the next season's flowering is almost entirely dependent on the growth made between the end of June
and the dormant period which sets in in the late
autumn. Long experience has shown that shading is
essential. This would hardly have been anticipated
from a knowledge of the fact that the habitat of the
ancestors of the Auricula is the sun-drenched slopes
of the Alps. Under our climatic conditions, however,
the foliage of both the wild and the cultivated kinds
bleaches to a yellowish-green colour and growth is
severely checked on exposure of the plants to intense
sunlight during periods of dryness. The necessary
shading is generally provided by standing the pots
on a bed of ashes on the north side of a wall, fence or
hedge. This is clearly a hit-or-miss procedure, for the
amount and intensity of sunlight is very variable.
Some measure of control can be obtained by placing
the plants in an open situation under movable
slatted screens. This involves more work but it has

the advantage that in dull weather the plants can be given the benefit of whatever sunlight there is. That this is considerable is shown by the fact that if the plants are kept in a position where no direct light reaches them their growth is poor and they either throw poor trusses or even fail to flower.

Whilst in their summer quarters, the ordinary routine of cultivation is lightened owing to its being no longer necessary to avoid washing the meal from the foliage when watering, although watering has to be done just as thoroughly as when the plants are under glass. The outbreaks of green fly which may be expected are easily controlled, for spraying with a dilute nicotine wash is no longer out of the question.

During this period, preferably when putting the plants outside, those which have not been re-potted may be given a small additional supply of food material. Until comparatively recently this was done by scraping off the top soil down to the level of the uppermost roots and replacing it with fresh compost. The procedure was abandoned on the ground that the roots did not penetrate upwards into the new soil. Now a light surface dressing of artificial manures has taken its place. Their satisfactory use requires judgment and observation, for it is easy to do more harm than good by over-estimating the quantities required. In these days when the potting shed contains sacks of various artificial manures it is easy to be too generous, with the result that the over-fed plants become coarse in their foliage and lose a good deal of the refinement of their flowers. Perhaps a little nonsense arithmetic may help to stress the fact that only the smallest quantities should be scratched into the surface soil. In agricultural practice from one to two cwt. of sulphate of ammonia per acre is the recognized dressing for several crops. Some indication of what the corresponding dressing per pot is can be roughly calculated as follows: an acre is 4,840 square yards and a hundredweight 112 lb, so one fortieth of an acre, that is, 121 square yards, receives a dressing of only about three pounds. About 50 pots can be packed on a square yard so on this basis three pounds of the manure would be sufficient for about 6,000 pots. This means a very small pinch of artificial manure per pot. But the dressing can be repeated at intervals of a few weeks, as the manure is leached out of the soil in pots, by the abundant waterings they receive, far more rapidly than in fields only exposed to rain.

Most growers who make use of artificial manures have their own special favourites and there are no investigations available to show which are the most satisfactory. But in horticultural practice there is now a distinct tendency to make use of the so-called organic manures rather than the chemically prepared artificials used by agriculturists. These include such substances as dried blood, finely ground hoof and horn, bone-meal, meat- and fish-meals, and, when obtainable, malt culms. Whether these materials are better than sulphate of ammonia, nitrate of soda, muriate of potash, etc., may be an open question but, as a group, they have the advantage of being comparatively safe to use. Malt culms, dried blood, meat- and fish-meals are rapidly acting, mainly nitrogenous manures, whilst hoof and horn and bone-meal only decompose slowly in the soil and gradually release a wide range of food materials. The grower will have to decide on which of these he will use and, remembering that Auriculas can be grown perfectly well on a good compost, he should limit the dose per pot to, say, as much as will cover a silver threepenny bit.

On the coming of autumn the plants have to be returned to their flowering quarters, partly to give them more light and partly to avoid the effects of heavy rain. Here they must be given all the ventilation possible by removing the lights from frames in dry weather and keeping all the green-house ventilators open. In wet weather the frame lights must be put back and ventilation provided by raising them from the rear. This continuous ventilation can be kept up throughout the winter for even if the soil in the pots is frozen into a solid mass it causes no damage to the plants. Watering has now to be reduced to a minimum, the plants requiring only a slightly damp soil. In extreme frosts, when the soil is frozen, no watering is possible and it must await the return of milder conditions.

The next operation is to tidy up the plants by removing dead foliage. The old leaves cease to function as the autumn wears on, turn yellow and die. They are most easily removed, just before they begin to dry up, by a cautious sideways pull. If left they decay whilst still attached to the plant and in doing so usually give rise to an abundant crop of the fungus *Botrytis* (p. 138). Once this is completed the plants are or should be reduced to firm cones of tightly packed leaves. The grower may then be happy in the knowledge that he has steered a successful course between the various imponderable factors he has had to deal with whilst cultivating his plants. His one anxiety now is lest sharp frosts may occur as the plants come into flower. Should they do so the trusses may be damaged. But at this time of the year the closing down of the lights or the ventilators is sufficient to ward off any ill effects.

The propagation of Auriculas has to be vegetative, for few of them come true to type when raised from seed. It is effected by means of cuttings or offsets. Some plants produce offsets so freely that four or five can be expected each season; others, even if they are large and sturdy, may only give rise to an offset once every two or three years. But however freely or sparsely offsets develop they are generally worth taking, since they can often be exchanged for other sorts if not wanted in the collection. The operation may be carried out at any time between the beginning of March and the end of August, but the most usual time is in June when the plants are being re-potted. After removing the soil immediately below the collar of the plant the offsets can be cautiously torn off with the finger and thumb or, if rather large, severed with a sharp knife, taking care that any roots they possess are preserved. The larger offsets will probably be well rooted and they may be put directly into 60s for next season's flowering. The smaller ones, with possibly a single root just pushing out, are best handled by setting them round the edges of a pot containing a sandy compost and then transferring them, when clearly well established, to separate small pots. Those which are unrooted will, almost invariably, quickly develop a root system in a compost consisting mainly of sand if it is kept moist and if transpiration is checked by covering the pot or pan with a bell jar. But small, unrooted pieces should not be taken late in the season; for, though they may root and show signs of growth, they often lose so much of their leaf surface during the winter that their chances of survival are slight.

The Auricula is not subject to the attacks of many insect or fungus pests and those to which it is liable can generally be kept under control without much difficulty. The one insect which will almost certainly be found on it at some time or other is the ubiquitous green fly. Its first appearance usually coincides with the time when the plants are under the most searching inspection, namely the flowering time; consequently its spread can be checked with the minimum of trouble. The first of the insects are generally noticed in the heart of the rosettes and a little later on the footstalks of the flowers, where they are clearly visible until the expanding flowers close up the truss. If seen in either of these positions the underside of the leaves of the infested plants should be examined. The simplest method of dealing with this early stage is to brush the insects over with a four per cent solution of nicotine. If the outbreak appears to be getting out of hand it must be controlled by means of a nicotine dust, for spraying is out of the question as it washes the meal from the flowers and foliage. Fortunately the dust, if uniformly and lightly applied, is almost invisible. There is no excuse for an early infection developing into an epidemic, but if this should happen, control is still possible by fumigating the frames or houses with tobacco shreds.

A second pest is the root aphid, *Pachypapella* (*Pentaphis*) *auriculae* Murray. This forms white woolly masses on the roots, often in such quantities that it is a surprise to find that the plant is still living. But though the plants tolerate it, their growth is appreciably checked. The first noticeable symptom of an attack is the formation of one of these woolly masses on the collar of the plant. This serves as a centre of infestation, and the insects composing it should be killed at once by brushing them over with methylated spirits. Their appearance is usually a good indication of a heavy root infestation. The trouble used to be kept under partial control by washing the roots with soap and water or a dilute nicotine solution when re-potting. This method of treatment, whilst certainly useful, could not guarantee that the last of the insects had been killed and that there was consequently no starting point for a fresh attack. Nowadays, however, the use of soil fumigants has brought the trouble well under control.

A most efficient insecticide, as far as root aphid is concerned, is paradichlorbenzene (P.D.C.B. for short), and it has the merit of being very simple to use. The material volatilises slowly, and if a piece the size of a pea is placed on top of the crocks, the vapour diffuses through the soil and soon does away with the trouble. This insecticide is also said to be effective against the vine weevil, the fat, creamy white, brown-headed, eyeless larvae of which feed on the roots and often completely gnaw through the upper part of the 'carrot'.

A disease known simply as the 'rot' was at one time the most serious trouble Auricula growers had

137

to contend with, and several harassing reports of its destroying large collections of plants are to be found in the literature of the early part of the nineteenth century. These accounts antedate the period when it was known that fungi could be responsible for plant diseases, so the identification of the casual agent is dependent entirely on the symptoms of the disease. These are similar to those produced by the grey mould *Botrytis cinerea* Pers. This mould is of almost universal occurrence during the late months of the autumn and if the dead foliage has not been removed from the plants a crop of its dark grey, fluffy spore-bearing tufts can be counted on to appear. The fungus is then growing saprophytically and doing no harm. It may, however, begin to attack the crown parasitically from the dead leaves, and once it has established itself on the living tissues of the plant the effects are devastating. It grows rapidly, killing all the tissues into which it penetrates, and the speedy death of the host plant is almost certain. Unless the plant is a treasured one it should be destroyed at once. If, however, its preservation is a matter of importance a drastic surgical operation offers the only chance of survival. The infected tissues must be completely excised even if this involves cutting off the whole crown of the plant. The exposed surface offers a favourable site for a fresh infection by air-borne spores, so it should be dusted with dry Bordeaux mixture or flowers of sulphur. Then there is a chance that an offset or two will develop on the old stump.

Little is heard now about severe epidemics of *Botrytis*, probably because of the greater attention given to ventilation and still more to the fact that over-feeding of the plants is not nearly so universal as it was. Nevertheless there are few growers who do not lose a few plants each season from the attacks of *Botrytis*.

The only other disease of any importance, and a very rare one, is that caused by the brown mould *Heterosporium auriculae* Mass. This appears as dark brown, powdery spots on the foliage soon after it has started into growth in the spring. It is unsightly but it causes little real damage to the plant. Dusting with dry Bordeaux mixture is said to be an efficient method of preventing its spread.

CHAPTER VII

AURICULA BREEDING

FROM the sixteenth century until the beginning of the nineteenth century Auricula breeding was entirely a matter of chance. The breeders simply marked down a variety which they considered to be an improvement on any then being grown, harvested its seeds, and with these raised a new culture in which they hoped to find still better kinds. In this long period thousands of amateurs and professionals must have raised millions of plants as the basis for their selections. The efficacy of the method can hardly be questioned, for by the latter date all of the groups of Auriculas now in cultivation had been produced: more in fact, since two, once popular, groups have been lost.

In the early part of the period the work was comparatively simple, because only the Alpine and Border groups, with their striped and double derivatives, were known. For the time, efforts were devoted in the main to securing larger and more richly coloured flowers. It was only later, when the Auricula had become a florist's flower, that the niceties of shape, the proportions of the various zones and colour

140

gradation were appreciated. The progress made in this limited field was naturally rapid and a half-century of selection was sufficient for the production of large numbers of plants which, in most respects, compared favourably with the present-day representatives of these two groups.

When the great break came and the Auricula acquired those unique characteristics, namely, the paste and the green edging of the flowers, pollination was still a haphazard process, depending altogether on the vagaries of honey-seeking bees. Yet, in spite of having to pay attention to these additional features, and, more important still, having to secure conformity with an elaborate set of points which defined the florist's ideal of a perfect flower, progress was made at a rate which now seems amazing; for the various groups of Show Auriculas were fully established in about half a century.

By the beginning of the nineteenth century it had become generally recognized that seed production was the result of a sexual process, and Koelreuter's experiments carried out on the Continent a generation earlier had shown the possibility of raising hybrids by controlled pollination. Auricula raisers were not slow in recognizing that this was likely to produce more definite results than the casual pollinations effected by bees. Their first method of control was to plant the sorts they required to interbreed in pans set out in an isolated position well away from

141

the main body of their Auriculas, thus still leaving the bees to carry out the rest of the operation. Emmerton, after testing out the method, advocated it with his usual enthusiasm and, what is more, insisted for the first time on the desirability of crossing like with like. To encourage others to take up the method, he published in 1815 lists of Green-edged, Grey-edged, White-edged, and kinds with a violet body colour, by crossing which he confidently promised his readers that they would raise 'some first-rate flowers, worthy of being named after the greatest heroes or beauties of ancient and modern times'. The next step was the obvious one of pollinating by hand and so securing greater control.

It can be assumed that these measures of control resulted in the production of comparatively simple mixtures of plants. But, even so, breeders looked upon the results as chaotic and only to be explained by saying that crossing 'broke the type'. However, they began to speculate upon what was happening and came to the conclusion that the direction in which the cross was made was chiefly responsible for the complexity of their results. A curious body of Auricula-breeding lore then came into existence. There does not appear to be any connected account of it and it was largely handed on by word of mouth, except by those who treasured the details of their system as the greatest of secrets. Possibly a fair summary of these beliefs obtained by rummaging among the early numbers of 'The Gardeners' Chronicle' is that the male parent determines the vigour of the plant and the colouring of its flowers, the female the shape of the flowers and their general refinement. It is hardly necessary to say that the system was not infallible.

In 1900 the rediscovery of Mendel's paper, 'Experiments in Plant Hybridisation' (1866), concerning the laws of inheritance of certain plant characters, completely altered the outlook of all concerned with plant breeding. Prior to this, as the botanist Lindley wrote, plant breeding was a gamble —with the odds on the plant. Then it became evident that law and order underlay the seeming chaos and that, given a knowledge of how individual characters were inherited, it was possible to predict accurately the results of crossing plants which carried them. Moreover it looked as if plant breeding could be systematized and a preconceived form of a new plant built up in the space of time demanded for the raising of two plant generations. The great improvements made with many economic crops during the past half-century bear witness to the correctness of these views.

But horticulturists as a whole, in spite of many examples proving its value, have little faith in Mendelism, and Auricula breeders generally flatly state that it has no application whatever to the solving of their particular problems. This view, held as it is by many breeders of great experience, obviously needs consideration. This requires, as a preliminary, an investigation into the manner of inheritance of as many of the features as possible which are of interest to the florist. But, so far, these problems do not seem to have attracted the attention of professional geneticists in spite of their interest in hybrids between distinct species and in the generally recognized complexity of the results of interbreeding the highly distinctive groups of Auricula now in cultivation. There is consequently only very limited information in the subject, and even this is often difficult to appreciate by those accustomed to thinking on Mendelian lines.

The simplest way to show the bearing of Mendelism on Auricula breeding is to describe the results obtained whilst attempting to follow out the story of the inheritance of a few important characters. Then the fundamental basis of the subject should become clear and the possibility of making use of Mendel's principles can be considered.

In making experiments of this kind it is essential to keep the problems as simple as possible by confining attention to a single pair of characters at a time. One such pair is provided by thrum and pin-eyed flowers, but there is the minor complication that there are two kinds of thrum, one breeding true to type, the other throwing pins as well as thrums. The true-breeding type can only be distinguished by raising a crop from self-pollinated flowers. This type of thrum when crossed by a pin, or vice-versa, gives a first generation consisting only of thrums. If some of these are intercrossed and a second generation raised, it consists of both thrums and pins, the former being the most numerous. This is explained by the fact that the reproductive cells, that is, the pollen grains and the egg cells, carry each either the thrum or the pin character—not both. Calling thrum A and pin a, the possible unions of the reproductive cells are:

POLLEN	EGG CELLS
A	A
a	a

The resulting plants have a constitution represented by AA, Aa, aA, and aa; aa will appear as a pin flower, and the others as thrums. Of these AA and aa must breed true to type when self-pollinated, as they carry only one of the pair of characters. Aa, on the other hand, is constitutionally similar to the first cross and if selfed it gives both pins and thrums.

The character which appears in the first cross and effectively masks its opposite number is described as 'dominant', and the one which is temporarily lost sight of as 'recessive', whilst the term 'heterozygous' is used to describe a plant with a constitution represented by Aa. Such a first cross is thus always heterozygous.

From the Auricula breeder's point of view the

145

main point to be noticed here is that the recessive character only appears when pollination results in the mating of one recessive unit with another, derived from a pin flower or from a thrum in which this character is latent. This information is often valuable. If, for instance, an otherwise good flower is spoiled by a pin eye it may still be of use as a parent, for the first cross with a pure thrum will consist of thrums only, since Aa is, as far as appearance goes, similar to AA.

Another simple example is provided by an experiment made with the object of breeding striped flowers (p. 71). The first cross made between a striped and a normally coloured Auricula consisted entirely of whole-coloured plants. A number of these were intercrossed in order to provide a large second generation. In this the striped character re-appeared in combination with several different colours. Here then striping proved to be a recessive character. The usual Auricula-breeding technique would have resulted in the loss of striping; for it almost invariably stops at the first generation, when, if the desired feature is not obtained, the plants are discarded.

A slightly more complex example is provided by a cross made between *Primula auricula ciliata* and a Self in order to obtain some information on the mode of inheritance of meal. In the former the leaves are small, free from any trace of meal, and the flowers

146

are yellow, whilst in the latter the foliage is large and mealy and the flowers almost black. In the first cross the foliage was a precise counterpart of that of *P. a. ciliata*, except that it was appreciably larger in size. The absence of meal is therefore a dominant character, but neither large or small foliage can be described as dominant or recessive. The heterozygous condition can be recognized here by the intermediate size of the leaves. The flower colour was reddish brown.

In a second generation raised from self-pollinated flowers the segregation and the recombination of these characters resulted in a mixed batch of plants, which, however, were easily sorted by going over them character by character. As far as meal was concerned they fell into two groups, one without and the other with meal, the former outnumbering the latter. Mealiness is therefore a recessive character.

The sorting of the leaves into small and large sizes, comparable with those of the parents, and intermediates, resembling those of the first cross, proved to be impossible, for they graded from small to large with no recognizable breaks in the continuity of the series. This lack of sharp segregation is characteristic of features associated with size, weight, shape, and so on. Such characters are described as 'quantitative'. They are common in the Auricula, and as the plants can be propagated vegetatively the watering-down effect of the intermediates is often valuable; it can bridge the gap between, say, too broad and too narrow a zone of body colour, or too much or too little meal.

The colours of the flowers of the second generation did not lend themselves to accurate grading or description but they could be classified roughly into three groups, one a constant yellow, one a reddish brown varying from a rich chestnut to burnt sienna, and the third, also variable, purplish black. In the original parents from which the first cross was made the yellow colour is associated with small meal-less leaves, and the dark colour with larger, mealy leaves. In the hybrid the reddish brown colour is associated with leaves of intermediate size. In the second generation, however, these associations broke down and yellow flowers, for instance, occurred in combination with large or small mealy foliage and dark flowers with small, meal-free leaves.

It is this recombination of the characters of the parent plants in the second generation of a cross which is the basis of most plant improvements. Unfortunately it involves growing on a large scale if every combination of more than three or four characters is to be obtained. In the first example in which only a single pair of characters was concerned there were three constitutionally different kinds, AA, Aa, aa; if another pair of characters such as freedom from meal M and mealiness m had also been taken into consideration, there would have been nine constitutionally different kinds, resulting from the combination of *MM*, *Mm*, *mm* with them, namely: *AAMM*, *AaMM*, *aaMM*, *AAMm*, *AaMm*, *aaMm*, *AAmm*, *Aamm*, *aamm*. A further pair of characters would give 27, and yet another, 81 combinations, the number increasing at too great a rate for anyone except a theorist to contemplate with equanimity.

Some of the difficulties of systematized Auricula breeding now become evident. In the case of the simpler groups, the original parental species and their immediate descendants, the Alpines and the Borders, they are no more troublesome than the ordinary run of flowers, simply because so few characters have to be taken into consideration. In the Edged groups, however, with their combination of floral and leafy characters producing the most complicated type of flower in existence the story is quite another one. There are too many plants to be handled unless growing plants by the thousand is a possibility. This is beyond the resources of most amateurs and professional growers nowadays, because a large percentage of the plants are not worth preserving.

The next difficulty is concerned with raising second generations. Many growers will dismiss it by saying that it takes too long for those who want to secure results quickly. But sometimes, as for

149

instance in the recovery of recessive characters, it is essential. The greatest difficulty of all, however, is that, particularly when dealing with the Edged group, there is rarely more than the vaguest of information about the genetical structure of the parent plants. A number of characters can be recognized by inspection but there are also hidden recessives ready to play their part when cross-breeding offers an opportunity. Thus a notebook record of, say, 'cross 143, splendid × magnificent' may mean something to the Auricula breeder but it has next to no real significance for him.

On the face of it then Auricula breeding on Mendelian lines is not likely to make any great appeal to those who simply want to raise a few plants well up to the standard of the best in cultivation, since it means turning a pleasant hobby into a time-robbing, laborious pursuit. They will therefore have to be content with present-day practice, and, knowing that for years past its employment has resulted in the production of many exquisite sorts, they will carry on confidently. This is dependent on the fact that, particularly amongst the Edged groups, most of the sorts are heterozygous for a number of characters.

Before beginning his experiment the breeder generally has a clear idea of the plant he wishes to obtain. For instance, he might like to combine the excellent paste of 'splendid' with the nicely propor-tioned body colour of 'magnificent', and accordingly pollinates one with the other. The resulting plants show many re-combinations of the various characters of the two parents and, even if the desired one does not occur, there are almost certain to be others worth saving. If the desired plant does not materialize, the breeder may then argue that the characters he is particularly interested in must be present amongst the medley of sorts the cross gave rise to and that they still ought to be obtainable in combination with one another. The assumption is correct but the difficulty is to know which should be intercrossed to secure it. The problem could be solved, but only by laborious cross-breeding experiments, so if the breeder wishes to pursue his quest further he can only chance mating the right two together. The probability of doing so is a remote one and, on the whole, there is something to be said for shutting down the experiment before reaching this stage and for being content with any good sorts which may have been found. The operation is a gamble pure and simple—and probably this gambling is one of the greatest attractions breeding work has to offer!

The value of like-to-like crosses was recognized at an early stage of controlled crossing and the experience accumulated during the past century has shown that it greatly simplifies cross-breeding. This is due to the fact that when the parents resemble one another closely there are relatively few characters to be taken into consideration. In the most stringent form of this outlook a Green-edge is only crossed with a Green-edge, or a gold-centred Alpine with another gold-centred Alpine; but if greater latitude is allowed, the crosses should be confined to their own particular groups such as Edged × Edged or Alpine × Alpine. In either case the limitation results in the avoidance of a great deal of waste, and this nowadays has become a matter of considerable importance both to amateurs and nurserymen.

These narrow crosses undoubtedly are sufficient to keep up the standard to which the Auricula attained, it may be, as long as a century ago. They do not, however, offer any obvious chances of further developments. Whether fanciers should be satisfied with what they have now and be content with marking time indefinitely is open to argument. The answer of the plant breeder familiar with the progress recently made in the study of his subject would almost certainly be that further improvements should be sought for. He would suggest that by now every conceivable grouping of the parent characters has been obtained and that it might be well worth while to seek fresh ones by going outside the limits set by the original parents. The material available happens to be especially favourable for work of this nature. Firstly there are several very distinctive forms of *P. auricula*, such as *P. a. Bauhini* and *P. a. ciliata*, which might take its place. Secondly that bugbear of the species hybridist, namely sterility of the crosses, need not be anticipated, and finally Alpine Primula species are notorious for the freedom with which they inter-breed.

It is probable that those interested in this group of Alpine Primulas have carried out experiments which throw some light on the possibilities hinted at above, but no records of them have been available to me. A preliminary and very incomplete investigation of the subject indicates, if no more, that it is worth pursuing by anyone who has the necessary time and resources at his disposal. The study began with a cross between *P. a. Bauhini* and a Green-edge, the genetical constitution of which was fairly well known from previous breeding experiments with it. The first cross showed an almost complete swamping of the characters of the Green-edge, and the almost uniform batch of seedlings might well have been labelled × *P. pubescens*. The second generation in turn was reminiscent of the seedlings derived from this so-called species. They showed the usual preponderance of plants with reddish-brown flowers and the usual extensive colour range. The size of the flowers, however, was indeterminate, ranging from that of the smaller-flowered parent *P. a. Bauhini* to that of the larger-flowered Green-edge. Amongst them were flowers of various colours which had sharply defined, narrow white edges similar to the wire edges of some kinds of carnations. But there was no sign of a green edge accompanying them; in fact no true Green-edge appeared in a culture of over 150 plants. This wire edge was due to the thickening of the cell walls of the margins of the petals, which was carried so far that it practically obliterated the cell cavity. The edges of the petals thus resemble the cartilaginous margins of the leaves of *P. a. Bauhini*.

Here then is a starting point for a new group of Auriculas. It is early to judge its value, and all that can be said is that some of the sorts with a rich violet ground, for instance, set off by a narrow thread of silver, or with a ground colour of imperial yellow or glowing chestnut, are quite attractive.

A second new feature was provided by the meal of some of the plants, which, instead of being white, was of a pure sulphur-yellow colour. This, overlying the green of the leaves, produced an original and rather pleasing colour effect. This same colouring has, by further crossing, been obtained in the paste of a few Green-edged sorts.

The wide colour range, too, offers an opportunity for bringing back into cultivation many shades of orange-brown, burnt sienna, chestnut, red, etc., with which the earliest growers of the Auricula must have been quite familiar but which are now rarely seen. Apart from this the smaller-sized flowers appeared to be worthy plants for the alpine garden, so, after potting off the plants required for continuing the investigation, the rest were set out to test the matter. They grew freely and the large silvery star-like masses of foliage and trusses sometimes carrying as many as 50 flowers made a brave show in the spring of 1946. But the drought of that year and the lack of labour to contend with it was too much for them and most of them died out.

A still wider cross, involving a distinct species instead of a closely related sub-species, was made between × *Primula decora* and 'Linda Pope'. The former belongs to the *P. hirsuta* end of the Auricula series, and the latter has all the characteristics of *P. marginata*. Thus × *P. decora* has rather coarse green foliage and its velvety flowers are of a deep blue violet colour. 'Linda Pope' on the other hand has neat rosettes of deeply toothed, silver-margined leaves and pale lavender-blue flowers each with a small lightly powdered centre. It is one of the daintiest and most popular Auriculas—if it may be called an Auricula—now in cultivation.

The cross had to made on × *P. decora*, for the ovary of 'Linda Pope' is abnormal and functionless. The flowers of the first generation were all of an intermediate blue colour, some on green and others on silvery rosettes, indicating that × *P. decora* was heterozygous with regard to meal characters. In the second generation there was the expected colour range from pale, through intermediate, to deep blue tones, any one of which was associated with either green or mealy foliage. From this a set of plants

155

with silvery leaves and flowers intermediate or dark blue in colour were picked out; apart from one or two with pin eyes they were worthy companions of the exquisite 'Linda Pope'. There is evidence then that the further development of the Auricula has not necessarily ceased and that, as time goes on, new additions to its already comprehensive range will be made.

Hand pollination of the flowers of the Auricula is a simple and reliable process if due precautions are taken. The one point on which attention has to be focussed is that no unwanted pollen must, under any circumstances, find its way to the stigma of the mother plant. To make certain of this the flower chosen as the female has to be emasculated whilst it is in the bud stage. This is effected by noting the position of its stigma and then cutting off the tube and the petals with a sharp pair of scissors. At this stage the stamens should not be mature and consequently they are got out of the way before they can work any mischief. An alternative method of splitting the corolla and then tearing it out is less satisfactory, as the stamens may break during the operation and possibly self-pollinate the flower. The pollen supply is obtained from stamens which have only recently split open. Such stamens will generally be found on flowers just beginning to open. One of the stamens should be picked out with a finely pointed pair of forceps and its dusty surface lightly brushed on the previously exposed stigma. The smear of pollen which this leaves on the stigma can generally be seen with the naked eye. After pollinating three or four flowers in a sufficiently mature condition on the truss the remainder are snipped off, the truss labelled with its record book number, and the plant placed under a bell jar, where it should remain for a couple of days. This is advisable, for otherwise a visiting bee may bring with it an unwanted supply of pollen from some other plant. Before making further crosses the forceps require to be sterilized by dipping them into methylated spirits. When self-pollination is to be effected the flowers should be allowed to open under a bell jar and a slender camel-hair brush pushed through the thrum until it comes into contact with the stigma. Again, the brush must be sterilized before it is used further.

Auricula seed ripens slowly and it is rarely ready to harvest before the middle of July. From the time when a yellowing of the pedicels and the capsules indicates that ripening is approaching a careful watch must be kept in order to harvest the seeds as soon as possible after the splitting of the capsules. If left for a day or two after this stage has been reached, the seeds detach themselves from the central placenta and the slightest vibration of the now rigid stem may scatter them.

One of the questions often discussed by growers is

whether the vigour of the Auricula is deteriorating. The subject does not lend itself to investigation, and as it is one of many about which it is difficult to find evidence one way or the other, the most to be hoped for is that a general impression, which may or not be correct, can be obtained. The usual complaint is that new Auriculas come and go all too quickly. But this does not necessarily imply that the constitution of the plant has broken down, for the failure may be due to the new kind not living up to its early promise or to its cultivation proving difficult. This, as all gardeners know, is an everyday occurrence with many kinds of plants.

Continuous vegetative propagation is frequently blamed for a suspected loss of vigour, and a comparison may be made with well known varieties of potatoes which have dropped out of cultivation during the critic's lifetime. But the comparison cannot be justified, for the deterioration of the potato is due to a number of diseases caused by viruses to which, as far as is known, the Auricula is not subject. Others blame their forebears, who, they state, by grossly over-feeding their plants ruined their constitution. There can be little doubt that by doing so they ran into trouble, but it does not follow that this was of more than a temporary nature. Since those days, too, the repeated starting up of new strains from seed and the continuous selection of the most vigorous plants must have gone far to eliminate any weakness this unnatural course might conceivably have given rise to.

The case must then be regarded as not proven so far as any slow deterioration of the Auricula is concerned. But there is a further complication to be faced. The assumed deterioration is usually associated with the Edged groups. These, as a whole, are less vigorous than the Alpines: a fact which becomes evident even in the seedling stage. Soon after germination has taken place some of the seedlings are generally seen to be distinctly stronger in growth than others and they retain this vigour throughout their life. Knowing this, many growers, with an abundance of seedlings at their disposal, do not trouble to grow on the weaker ones even though it is commonly believed that these often give rise to plants with particularly good blooms.

Whilst the above facts are well enough established it is difficult to account for them. It may be that the change over from the normal to the leafy corolla has some connection with them, for such mutations may bring with them other deep-seated changes. They involve changes in the nuclear make-up of the plants, and perhaps the symmetrical arrangement of the chromosomes, on which the familiar segregation and recombination of characters is dependent, is upset. Some nuclei may possibly have fewer than their normal complement of chromosomes, and analogy with other plants suggests that such plants would be

159

lacking in vigour. This, however, is only a speculation which may some day tempt a cytologist, whose business it is to investigate such matters, to look into the problem. But whether its solution will be of any help to the florist is an open question. More interesting to him is the expectation of life of a new sort of Auricula. Like all other plants it must wear out and perish through sheer old age, and even vegetative propagation cannot guarantee its immortality, for each new plant resulting from this is but a portion of the original plant. The length of life is very variable and dependent almost entirely on the environment provided by the grower. If it meets the requirements of the plants Auriculas are singularly long-lived. Even today, sorts distributed for the first time well before this century began compete on level terms with the newest productions, especially at the shows of the Northern Section of the National Auricula and Primula Society. Foremost among these is 'George Lightbody', raised by Headley in 1857, which, in spite of its ninety odd years, is one of the best Grey-edges in existence. Almost its equal in age and merit is 'George Rudd'. 'Colonel Champneys', raised in 1867, was usually obtainable in the immediate pre-war years from the firm of James Douglas, and, though a little chary in its production of offsets, still throws better trusses of violet flowers than any of its more modern rivals. It is probable that even these veterans are young compared with some of the cottagers' Alpines and such old-world sorts as the 'Dusty Millers'. But their history is lost. Even though only comparatively few sorts have survived so long, their existence is an encouragement to the present-day breeder; for the pleasure they have given to so many, over so long a period, is one of the things in life which is worth working for—if indeed work is the right description of a delightful hobby.

ADDITIONAL REFERENCES

BIFFEN, R. H. (1939). Notes on the development of Auriculas. *National Auricula and Primula Society* (*Southern Section*), *Annual Report for 1938*.

—— (1942). The development of the Auricula. *Journ. Roy. Hort. Soc.*, LXVII, 187.

—— (1950). Meal. *National Auricula and Primula Society* (*Northern Section*) *Year Book, 1949-50*.

CHURCHILL, G. C. (1886). Origin of the Auricula. *Gardeners' Chronicle*, XXV, 563. (This is a summary translation, with comments, of Kerner's paper, 'Die Geschichte der Aurikel).

ERNST, A. (1933). Weitere Untersuchungen zur Phänanalyse, zum Fertilitätsproblem und zur Genetik heterostyler Primeln. I. *Primula viscosa* All. *Archiv der Julius Klaus-Stiftung, Zurich*, VIII, 1.

—— (1936). Ditto. 2. *Primula hortensis* Wettstein. *Archiv der Julius Klaus-Stiftung, Zurich*, XI, 1.

—— (1938). Ditto. 3. Die F_1-Bastarde *Primula* (*hortensis* × *viscosa*). *Archiv der Julius Klaus-Stiftung, Zurich*, XIII, 1.

—— (1940). Entstehung, Erscheinungsform und Fortpflanzung des Artbastardes *Primula* (*Auricula* L. × *viscosa* All.). *Archiv der Julius Klaus-Stiftung, Zurich*, XV, 1.

162

ADDITIONAL REFERENCES

—— and MOSER, F. (1926). Entstehung, Erscheinungsform und Fortpflanzung des Artbastardes *Primula
pubescens* Jacq. (*P. auricula* L. × *P. hirsuta* All.) *Archiv
der Julius Klaus-Stiftung, Zurich*, I, 273.

FARRER, R. (1913). Primula hybrids in Nature. Report of
the Primula Conference, 1913. *Journ. Roy. Hort. Soc.*,
XXXIX, 112.

HAY, T. (1940). Hogg and Emmerton on the Auricula. *Journ.
Roy. Hort. Soc.*, LXV, 13.

HIBBERD, S. (1886). On the origin and history of the Florists'
Auricula. *Journ. Roy. Hort. Soc.*, VII, 191.

KERNER, A. (1871). Können aus Bastarden Arten werden?
Oesterreich. bot. Zeitschr., XXI, 34.

MACWATT, J. (1913). European Primulas. Report of the
Primula Conference, 1913. *Journ. Roy. Hort. Soc.*,
XXXIX, 103.

LOTSY, J. P. (1925). Studien an wilden Bastarden zwischen
verschiedenen Linneonten. *Genetica*, VII, 220.

MORETON, C. Oscar (1949). The rise and development of the
Edged Auricula. *National Auricula and Primula Society
(Southern Section), Annual Report for 1948.*

SCHROETER, C. (1926). *Das Pflanzenleben der Alpen.* Zurich.
pp. 804-5.

STRUB, W. (1940). Untersuchungen zur Phänanalyse und
Cytologie des Artbastardes *Primula (Auricula* L. ×
viscosa All.). *Archiv der Julius Klaus-Stiftung, Zurich*,
XV, 105.

WETTSTEIN, R. (1898). Anton Kerner von Marilaun. *Ber. d.
deut. bot. Ges.*, XVI (43).

—— (1920). Die Verwertung der Mendelschen Spaltungs-
gesetze für die Deutung von Artbastarden. *Zeitsch. f.
ind. Abstammungs- u. Vererbungslehre*, XXIII, 200.

WIDMER, E. (1891). *Die europäischen Arten der Gattung Primula*. Munich.

WRIGHT SMITH, W. and FLETCHER, H. R. (1948). The genus Primula: Sections Cuneifolia, Floribundae, Parryi, and Auricula. *Trans. Roy. Soc. Edinburgh*, LXI, 631. (This paper contains many references to the Garden Auricula and to the species of Primula related to it.)

PLATE I

Crimson and white striping. The flower is a Self, and the
white stripes are due to meal. (Enlarged)

PLATE II

Scarlet, yellow and white striping on a Self flower. (Enlarged)

PLATE III

Grey Edge with a well defined silver margin and a poor,
granulated paste. (Enlarged)

PLATE IV

Green Edge with no trace of meal. *Ex* Virescent cross. (Enlarged)

PLATE V

A silvery Grey Edge with only a trace of meal.
Ex Virescent cross. (Enlarged)

PLATE VI

Grey Edge without a distinct silvery margin. Body colour
vandyke brown to gold. *Ex* Virescent cross. (Enlarged)

PLATE VII

Willow green Grey Edge, body colour crimson purple. Grown in an open border, and the flowers thinned down to about half their original number. *Ex* Virescent cross. (Enlarged)

29889231R00106

Printed in Great
Britain
by Amazon